A CHARTWELL-BRATT STU

Fourier Transforms in Action

by Frank Pettit
MA(Oxon) C.Eng, MIERE, MBCS

Chartwell-Bratt   **Studentlitteratur**

©Chartwell-Bratt (Publishing and Training) Limited, 1985.
ISBN 0-86238-088-X

Studentlitteratur
ISBN 91-44-24531-9

Printed in Sweden by Studentlitteratur, Lund

Contents

List of Figures

Preface

This book is intended for people whose 'first' subject is not mathematics.

The Digital (or Discrete) Fourier Transform is one of many analytical algorithms used in data analysis. It is a member of a group of techniques known as Harmonic Methods. These methods attempt to describe certain forms of time-series (and other) data in terms of frequencies - especially in terms of sets of harmonically related frequencies.

The book arose from courses developed and given at the University of Oxford since 1974. The courses arose from the 'needs' of research workers in many disciplines to use the popular Fast Fourier Transform (FFT) technique in laboratory data analysis. Such 'needs' are sometimes created by little more than fashion. The widespread availability of the FFT packages in many Mainframe Libraries around the World gave impetus to their practical usage. Once the trend was initiated and FFT results were used in reports and theses, the use of FFTs spread rapidly.

Sadly, not all applications of the FFT technique could stand up to rigorous analysis, and in a number of instances 'effects' attributed to features of the data were, in reality, due to the special characteristics of the FFT technique.

The Oxford courses were thus designed to provide an adequate practical grounding in the characteristics of the FFT but without involving the research workers in a rather gruelling mathematical formalism. Rather, from the formalism was derived a graphical representation embedded in a special purpose software package called 'HARMANY' (HARMonic ANalYsis). The mathematical formalism was thus translated into an equivalent but more 'accessible' formalism via an interactive graphical demonstration and study package.

The particular features of the Fourier Transform system which can give rise to difficulties of interpretation (inference) of the graphs were deliberately exaggerated by the careful choice of a rather coarse spectrum which also suited lecture theatre TV displays.

Most of the illustrations used in this book are taken from the videographics of the HARMANY package. They are

presented in a sequence which seeks to develop the concepts
leading to the more effective use of FFT packages. However,
some other items of software were produced for this book. One
such package illustrates initial data access by sampling
whilst another adopts an FFT algorithm used on both audio and
biomedical recordings.

When the subject matter of this book has been digested,
recourse should be made to more conventional works on the
subject of Signal Processing and to the descriptions of FFT
algorithms available in computer software libraries.

F.R.Pettit, M.A., Chartered Engineer

September 1985

Chapter 1

About Signals

1.1 Types of signal

In this work, we shall deal with 'single-channel', single-dimension signals. These will normally be signals varying in time such as speech, but they could equally well be variations in a spacial or a frequency dimension. An example would be the spacial distribution of stress in a structure or plate under test. Extension of the concepts to 2-D or 3-D represents, so far as computer analysis is concerned, merely an increase in workload and not a new technology.

Given the 1-D signal, varying in time, we can classify a number of 'types' of such signal. There are signals which repeat cyclicly such as the movement of a piston in a cylinder. There are signals such as TV scans which, whilst being single-channel, represent sequences of 2-D images. The TV signal consists of a 'message frame' (synchronization pulses and the 'colour burst') plus a dynamic field of intensity (luminance and chrominance) data. There are signals such as speech where repetition is rare. There are 'events' which by their nature are not cyclicly (if at all) repetitive. With any of these signals there can be 'noise'.

Clearly then these types of signal appear to require radically different processing techniques. We shall find that we can sometimes be tricked into use of a method which, whilst appearing good, can let us down. Sometimes we can justify the 'curious' use of a seemingly inappropriate technique.

1.2 Stationary signals

A signal is said to be 'stationary' when it appears to repeat in a regular pattern.

We lock the trace of a sine wave on an oscilloscope screen, using a linear time-base for x-sweep. The resulting display is stationary. The signal IS a variation in time - our use of a repetitive time base of reference imposes the stationary appearance of the image.

Fig. 1.1 illustrates a group of 'special' signals which will be considered in detail later. The upper two traces are clearly 'stationary'. The third trace is of an impulse and we must query whether this is a singularity or whether what we show is a synchronously repetitive sequence of such pulses. The lower trace is non-stationary in that the trace is simply typical of an ongoing, ever changing stream of variable-duration pulses.

1.3 Irregular signals

Speech and music are clearly non-stationary. Their irregularity of waveform clearly requires a different form of treatment than would be used for 'stationary' waves.

We include amongst the irregular signals, ECG traces. Fig. 1.2 gives the appearance of being stationary, but, a little further along the recorded tape from which these traces were taken we see Fig. 1.3 where, presumably the contact pad had moved just at the time when a heartbeat occurred. In Fig. 1.4, we see constant spacing of waves whose amplitudes vary.

1.4 Events

Little is to be found in signal processing theory on either the classification or the processing of 'events', though they form a major group of 'interesting' signals.

An ECG of heartbeats is frequently treated as though it represented a stationary signal. Cardiac operation is non-synchronous, each 'beat' being an event whose occurrence is dependent upon many factors. For many years, medical

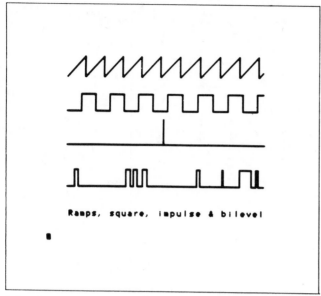

Ramps, square, impulse & bilevel

1.1   Some rather special signals

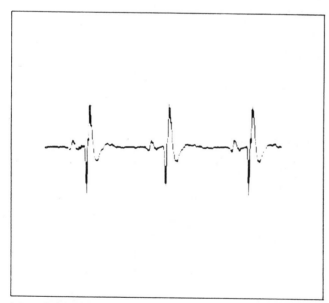

1.2   A repetitive wave?

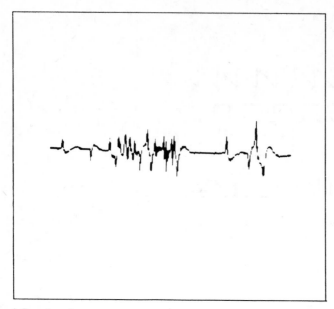

1.3   A noisy wave

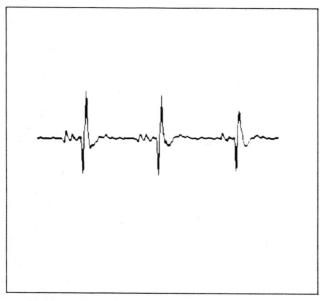

1.4   A not so repetitive wave

literature carried reports of work which treated ECG signals as though they were stationary. Much of the work was mathematically invalid, being based upon dubious assumptions.

Most oscilloscopes include a timebase trigger facility to permit the 'capture' of an event at a random point in time. Such behaviour may be copied in digital work.

The sequence of events in Fig. 1.5 looks innocent. However, add some noise as in Fig. 1.6 and the picture changes dramatically. There would appear to be an additional event which is in fact, just a noise spike.

Disturbances of this nature will commonly need to be eliminated prior to regular analysis.

1.5 Noise

Noise may be considered to be a 'random signal'. Originating in physical systems, noise may display various characteristics. Noise may occur in 'bursts' (events?), it may cover a wide frequency band or a narrow frequency band. It may be systematic - such as 'ignition interference' with broadcast reception, or the characteristic 'bars' of 'noise' on a TV due to the local use of a commutator motor such as a floor cleaner.

Noise is commonly an unavoidable element in the acquisition of a signal. The sensing of a foetal heartbeat carries a number of different electrical 'noises' such as the maternal heartbeat and muscular noise due to breathing and other movements.

In signal processing we must strive to acquire 'clean', minimal-noise signals but always allow for the occurrence of either 'bursts' or 'spikes' of interfering noise.

Fig. 1.7 illustrates a common form of noise. This is systematic and of varying magnitude. The frame shows 'bipolar' noise. Frequently such noise moves only 'positively' from a zero baseline (unipolar).

NOISE SHOULD NOT BE UNEXPECTED. A signal analysis

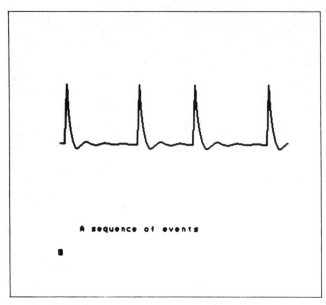

A sequence of events

1.5   Some 'events'

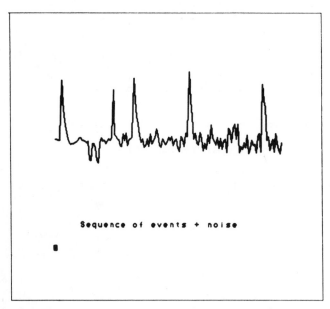

Sequence of events + noise

1.6   Some noisy events

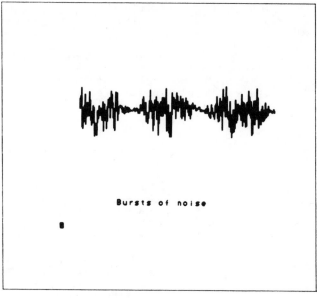

1.7 Just systematic noise

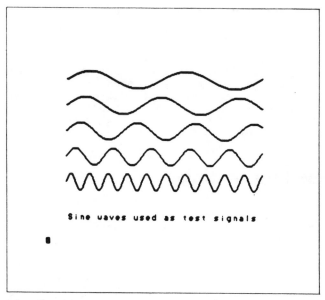

1.8 Test signals (1) sine waves

package must provide noise-handling facilities. There can be little sympathy for a researcher whose results are invalidated by 'unexpected noise'.

A medical research team working on knee defects first tried use of sensitive microphones strapped to the knee of a patient. Room noises were sensed and could have prevented analysis. Accellerometers were substituted for the microphones - result, clean signals and fine analyses.

1.6 Data errors

In both analog and digital work we must expect data errors of many types. There is the uncertainty resulting from a varying number of days in a month which makes the figures for one month appear better or worse than those for another month.

There are many examples of this 'time base jitter'. A writing tablet designed to digitize pen movements used a pen with a minute 'spark plug' at its tip. The sparks were initiated from a crystal oscillator to ensure stability. The resulting data showed 'noise' values some hundred times the magnitude of the 'signal'. The designers of the tablet had forgotten that a spark leaves the local air in an excited state leading to early ignition of the next spark. The subsequent spark is then 'late' - leading to the 'early late syndrome'. This type of noise can often be 'tamed' despite its magnitude.

There are also gross errors (rogues) due to induced electrical or magnetic pulses or to a dropped bit in a data stream. Such errors must be handled quite differently from time-base errors or other forms of noise. In practice, such errors and rogues must be hunted out prior to transformation, processing and the drawing of conclusions.

1.7 Signals used for testing

To test the effectiveness of a transformation or

processor algorithm, one uses synthetic 'test signals'. These take various forms, each with its own characteristics.

Fig.1.8 shows a group of simple sine waves of various frequencies but with a common, fixed amplitude. With these one may ensure that the Transfer Function is such that the processor responds as required to such waves.

In Fig. 1.9, another group of test signals is illustrated. The transforms of most of these waves are examined later in the book. Each has a transform of known characteristics and thus may be used in assessment of the performance of an algorithm under development or test.

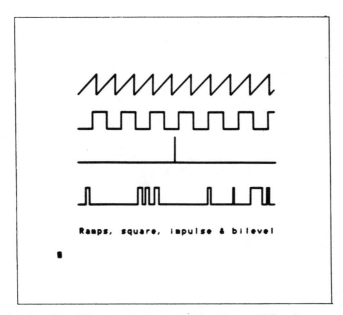

Ramps, square, impulse & bilevel

1.9   Test (2) ramps, rectangular, impulse and bilevels

Chapter 2

Sampling Techniques

2.1 Data acquisition

To perform digital analysis one must acquire a digital version of the signal. Rather than a continuous indication by some physical parameter, we must obtain a sequence of numbers representing that parameter.

Clearly, a digital signal is an approximation.

There are two aspects to this approximation:

1. There must be a sequence of 'snapshot values' taken over a period of time. Too many 'samples' result in an unnecessary storage and processing load whilst insufficient samples give wrong answers.

2. Each sampled value will be represented as a number of 'digits'. We 'think decimal' and thus have a 'feel' for three, four, five etc. decimal digits as being 'adequate' or, to be more precise, 'necessary and sufficient'. A digital processor commonly operates in binary and thus is designed to deliver six, eight, ten, twelve or 16 bits. Adequate, but not excessive, digitizing is the aim.

2.2 Windows

When we take a sequence of samples, we have effectively opened a 'window' onto a world which contains (only) the signal. For all we know, time commenced infinitely far to the left and extends infinitely far to the right of the window. Within that window-in-time we have a number of samples.

There are two possibilities when we 'see' a single wave

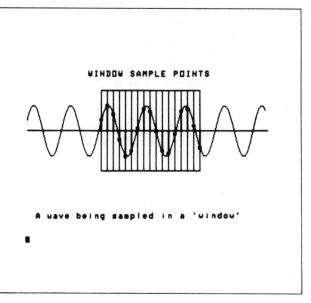

2.1 Sampling a wave in the window

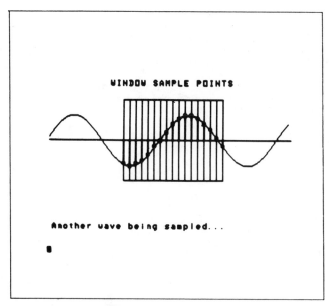

2.2 Sampling another wave

in the window.   One is that this is a unique event in time,
the other is that a precisely similar wave existed in all
preceeding windows and will occur in all successive windows.
The most usual interpretation of the underlying mathematical
theory is that our working window contains a wave which would
be repeated in all preceeding and in all succeeding windows.
In other words, we have picked one representative cycle of a
repetitive wave.   That this is clearly not the normal
practical situation will be demonstrated.

        Fig. 2.1 demonstrates the sampling of a 'continuous'
wave.   The sampling did not commence at the zero crossing
point of the wave and sampling ended on an incomplete number
of wave cycles.

        Fig. 2.2 shows a similar situation involving a different
wavelength.   These two simple examples indicate the
improbability of simply sampling a passing wave and expecting
to trap an integral quantity of 'cycles'.

        In Fig. 2.3 we have removed the graphical aspect of the
window 'frame', leaving a graph of the samples which were
drawn as the wave passed by.   The amplitudes of these are the
data points, and we are assuming that the time intervals
between sample points are invariant.   One common source of
data error lies in irregularity of sampling interval.

        Taking the data of Fig.   2.3, we have used a Fourier
Transform to produce the spectrum illustrated in Fig. 2.4.

        Rather than as we may have expected, a single spectrum
line showing the single frequency of the wave, the spectrum
contains many harmonics having both sine and cosine
components.

        However, just to show the consistency of the system, we
have taken an Inverse Fourier Transform which has enabled us
to reconstruct the sampled wave in Fig. 2.5.

        Interpretation of Fourier spectra is the principal
purpose of this study.

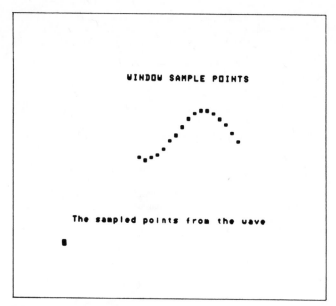

2.3   The stored samples

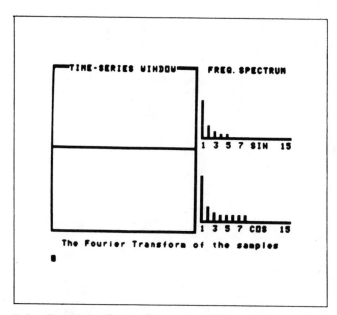

2.4   The Digital Fourier Spectrum of the samples

2.3 Sampling

To obtain a sample from a varying signal, we must take our sample as rapidly as possible. For infinitely short sample time, a precise mathematical theory of sampling has been developed. In reality, we must obtain energy when we sample the signal thus we must use finite time for the sampling act. The mechanism of sampling must then be an approximation.

Once the sample of energy is obtained, it must be digitized - this again takes time. To ensure that we can digitize the sample we need to hold the analog value in a special circuit during the process of digitization. Hence the need for a Sample and Hold circuit. The digitized numeric value must be stored in a register to await transfer to the digital processor (usually a computer).

2.3.1. How many samples?

At least twice as many as the highest 'wanted' frequency, but not more than say, ten times that quantity. This raises yet another enigma. What if the wanted aspect of the signal is spoiled by frequencies very much higher? These must be removed by a low-pass filter prior to sampling in a process known as 'signal conditioning'.

Similarly, if the signal contains frequencies lower than that of the 'window', these must be removed by a high-pass filter as another part of the signal conditioning process.

2.4 Digitizing

How many bits should we acquire? Sufficient to cover the 'dynamic range' (loud-to-soft) of the signal adequately. With special precautions, it is possible to convert speech to digital form and reproduce it understandably using just eight

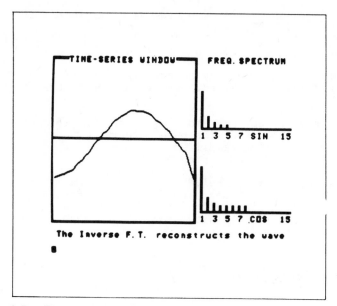

2.5 The wave reconstructed by Inverse Transform

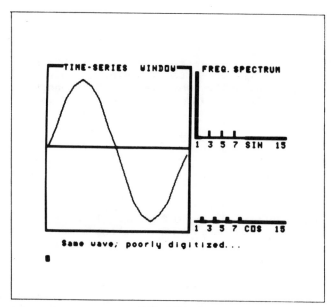

2.6 A 'pure' wave and its Spectrum

bits.   For good speech quality some twelve to sixteen bits
give reasonable results.

For fine ECG work, eight or ten bits is often adequate.

For image processing, much can be achieved with from one
to eight bits.   High-quality TV requires around ten to twelve
bits.

Fig. 2.6 shows an effect due to inadequate digitizing
(insufficient bits).   The otherwise pure integral wave has a
spectrum which contains low amplitude harmonics at even
positions in the cosine and at odd sine positions.

Signals may be 'compressed', e.g. in a non-linear
processor to restrict the 'contrast', and subsequently
'expanded'   back   to   'normal'   after   processing   or
transmission.   'Compander' circuits for this work are
available in microchip form.

2.5 Practical difficulties

2.5.1. Speed of conversion

Many types of Analog-to-Digital Converters (ADC) are
available ranging from milliseconds to sub-microseconds in
total conversion time, and with a variety of bit-counts.
Cost increases with number of bits and with higher conversion
speed.

2.5.2. Alias frequencies

Given a window containing say, 10 identical waves, ten
equally-spaced samples will obviously each give the same value

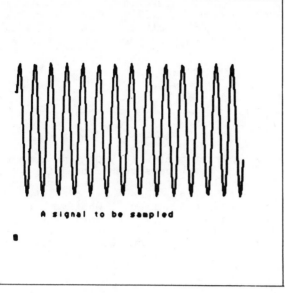

A signal to be sampled

2.7   A wave to be sampled

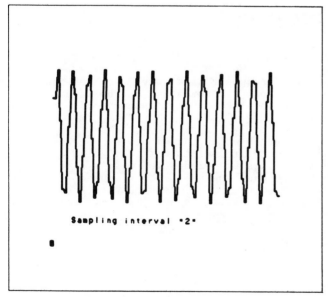

Sampling interval "2"

2.8   Reconstruction from sample interval 2

- effectively a DC 'signal', information has been lost.
Taking samples less frequently will result in alias
frequencies, that is sets of values which can appear to
represent waves of lower frequency than that of the sampled
signal.

2.5.3. For you to do

Try the following experiment with your home computer.
Write a program to set up a graphics screen containing 20 sine
waves, storing their 200 values (ten values per sine wave) in
an array. Select every 5th array element and plot the value
at its correct position on the screen. What you see is an
ALIAS frequency, lower than the actual frequency of the source
wave. Try selecting (sampling) at different intervals along
the 200-element array and observe the curious patterns which
result.

A selection of the types of result you may expect to
find are illustrated in Fig. 2.7 to 2.10. Fig. 2.7 depicts a
wave, the following illustrations are of successively coarser
sampling (greater sampling intervals).

2.6 What is 'frequency'?

In the real world, frequency of a wave is 'obvious'.
Once sampled, we prepare for transformation of the wave in the
window from the time domain into the frequency domain for
analysis.

The longest wavelength which the window can accommodate
is one which occupies the entire width of the window. A
signal which precisely fits the window and displays just one
'cycle' is fundamental to the analysis of frequencies. A
single sine wave which precisely fits the window is known as a
'first harmonic', two such waves in the window form the
'second harmonic'.

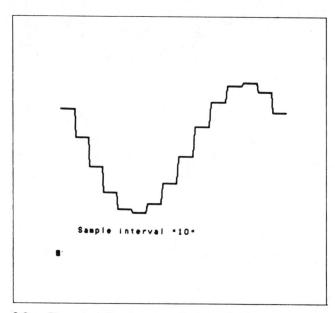

2.9   Reconstruction from sample interval 10

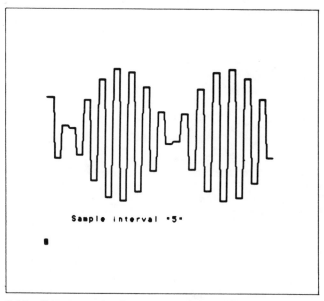

2.10   Reconstruction from sample interval 5

By adding together a number of such harmonics at particular amplitudes we can simulate 'any' wave. The converse is that we should be able to take 'any' wave and express (or describe) it as a set of harmonics of the fundamental window 'frequency'. Thus we build up the concept of Harmonic Analysis.

In Harmonic Analysis then, a 'harmonic' is related INDIRECTLY to the source signal via the sampling window.

In the 'real time' waves shown in the chapter on Fast Fourier Transforms, the window had a time duration of 4 milliseconds during which 256 samples were drawn from a selected analog source. The fundamental source wave frequency (giving one cycle in the window) was thus 25 Hertz.

Given a source wave of 100 Hz, our (25 Hz) window would display 4 complete cycles - a 4th harmonic spectrum bar. Given that our 100 Hz wave contains both 2nd and 3rd harmonic components, these would appear at the 8th and 12th spectral bar positions.

We tend to use the term 'frequency' all too easily in discussions related to harmonic analysis. What we usually mean is 'harmonic number' whose value should be multiplied by the window frequency to obtain equivalent source harmonic frequencies. We must however, not lose sight of the fact that the harmonics are frequencies of analysis and may have no existence in the real world.

Chapter 3

Fourier Transforms

3.1 Time domains and frequency domains

When a series of numeric values from samples of a signal is taken 'in time', the values and operations on them are considered to be 'in the time domain'.

When a signal is viewed as a set of frequencies of particular phases and amplitudes, the values and operations on them are 'in the frequency domain'.

When designing an 'electronic organ', we use a set of Tone Generators and operate 'in the frequency domain', finally adding Envelope Shaping 'in the time domain'.

Pitch of an organ note is set against a 'natural' scale in which an Octave is divided into twelve 'semitone' frequency intervals. The 'octave' comes about by using eight of the twelve 'notes' as in some way 'dominant' and calling the others by such terms as 'accidentals' (or "sharps & flats").

There is a Harmonic Relationship between certain tones - C & C' have a 1:2 frequency relationship. C & G have the ratio 2:3.

3.1.1. For you to do

Set out the names of the piano notes... C, C-sharp, D etc. Mark out the interval numbers (C-sharp = 1, D = 2 etc.). Note that G is at the 7th interval and G' is at the 19th. Define the frequency of G' to be 3 times that of C. Now count up 19 from C (to G') and divide by 2 (count down 12 intervals) to reach G. Repeat this process from G (up 19, down 12, i.e. up 7) but, to reach the range C to C' count down another 12 (divide by 2 again). Continue this process until it repeats when you reach C again. You will have multiplied by 3 twelve times and divided by 2 nineteen times (to get from C to C'). However the ratio of these two numbers indicates an error of just over 1%. How then do you tune a piano or an organ? The answer is that you cannot - other than using an accepted sequence of approximations using other ratios than 3:2. Some aspects of music use harmonic ratios - others provide 'rich' near-harmonic ratios. Use of multiple strings or pipes which are deliberately detuned by one or two Hz improves the richness of timbre.

3.2 The timbre of a musical note

Timbre of a musical note is determined by the existence of 'harmonics', 'near harmonics' and heterodyne beats of the 'fundamental' tone.

Thus, the physics and mathematical analyses of musical sounds involves time series, harmonic and non-harmonic relationships. Musical works also involve other 'beats' forming regular and irregular patterns.

We may use mathematical analyses either in the time or the frequency domain, transforming as required from one to the other.

3.3 Continuous and discrete harmonic transforms

By far the most popular of the mathematical techniques of frequency analysis is that group of transformations and operations associated with Fourier Analysis. This technique transforms WHATEVER time domain waveform is present into a HARMONIC SERIES whether the source data involved harmonics or not!

A Fourier Transform (FT) of a pure note C may be a

single spectral bar.   Its corresponding C' would be a single
spectral bar at the next harmonic position.   However the
associated note D would appear as a very full spectrum of
harmonics   - the harmonics of a particular mathematical
analysis.   There are no 'in-between' positions in a harmonic
analysis - just a 'fundamental' (e.g. a wave which just fits
the width of the window) and a series of harmonics (2nd, 3rd
etc.) up to 'infinity' (or in digital work, up to half the
number of samples in the window).

     Thus we operate as though the only frequencies available
to us are 1, 2, 3, 4 and so on.   There is no such harmonic
frequency as 3.14159, but we shall see later what the spectrum
of such a wave is like.

     The 'Fourier Series' contains an infinite number of
terms (harmonics) and was intended for the analysis of 'waves'
in the generalized sense - 'continuous waves'.

     For the purpose of Harmonic Analysis, time began in the
'infinite past' and will continue into the 'infinite
future'.   An 'event' is the coming together of an infinite
set of harmonics which cancel for most of eternity - just
entering a specific phase relationship so as to 'appear' (as a
pulse) for a brief while.

     The converse, the analysis of an 'impulse' is well
understood as an infinite harmonic series.

     Thus, there exists the Continuous (or infinite) FT and
of course, its inverse, the IFT - a formula for transforming a
Fourier spectrum to a time-series wave.

     The   FTs   are   LINEAR transforms.   If you take a
transform, scale the results and invert, a scaled version of
the source wave is obtained.

     THE FT IS AN ALTERNATIVE DESCRIPTION OF THE DATA.   We
could view '3' and 'three' as transforms - alternative
'descriptions' of a concept.   The one representation is best
suited to certain processes whilst the other is better suited
to other forms of process and communication.   The FT is a
TRANSFORM, it is not a FUNCTION.   It provides an ALTERNATIVE
DESCRIPTION of the wave.

3.4 Discrete Fourier transforms

For digital processing, we work with a set of samples drawn from a window on a (continuous) wave. The formulae are converted from 'continuous integrals' to 'discrete summations'.

In this conversion from FTs to DFTs, we lose something. We lose precision of analysis. We gain something. We gain the ease of digital processing.

In use of DFTs we must ensure that what we lose does not invalidate the work and that what we gain is not outweighed by excessive and expensive processing time.

3.5 Time series and frequency spectra

We make extensive use of graphics in time series and Fourier Spectrum work.

The time series may be represented with time as the X-axis and wave amplitude as the Y-ordinate. the Y-ordinates would be joined by lines to simulate the (continuous) source wave. Brightness or colour may be used to discriminate between 'raw' and 'processed' data.

The spectrum may be represented with frequency as the X-axis and harmonic amplitude as bars in the Y-axis. Phase of harmonics is most readily represented analytically by use of two spectra - one displaying a SINE analysis, the other displaying a COSINE analysis. Relative amplitudes of the sine & cosine components of one harmonic 'gives' the phase relationship of that harmonic in relation to the window.

Such a spectrum is called a COMPLEX SPECTRUM - and is virtually the standard technique.

Two examples of such waves and their spectra are given in Fig. 3.1 and 3.2. The first uses an 'ordinary' DFT operating on a window of just 16 samples. The second wave is

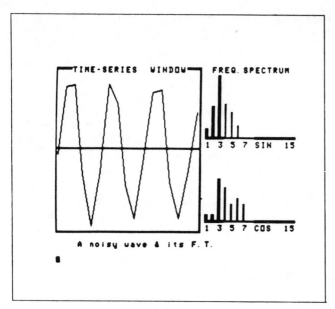

3.1   A noisy wave and its DFT

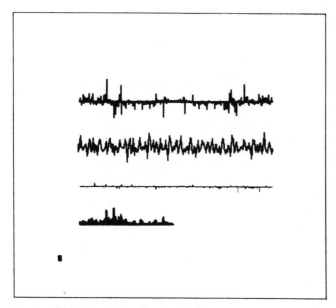

3.2   'FFT' of a cathedral organ

sampled at 256 points and a FFT was used to provide the cosine spectrum shown above the time series graph and the sine spectrum below.

To achieve relatively high resolution in analysis we need many samples from which to derive many harmonics - 2048 & 1024 are common.

The digital summations of many sine & cosine functions consume considerable computer time. Because the same wave angles appear many times, we can economize by computing a set of sine & cosine values once only. By sorting the time-series values in a particular manner further time-efficiency is achieved.

The most popular of these techniques is due to Cooley & Tukey and is known simply as the Fast Fourier Transform - the FFT. There is a corresponding inverse - the IFFT (or FIFT). These are by implication, DFTs - but the terms FDIFT etc. have thankfully, not come into use. Many varieties of FFT algorithm have been developed and the great similarity between the FFT and the FIFT in some designs has rendered a separate FIFT algorithm unnecessary. A single algorithm performs both forward and inverse transforms. The FFTs are now considered to belong to the general class of Z-transforms.

3.6 The formulae

Just for the sake of convention we should spend a moment with the mathematics of the Fourier Transform. First, a look at the Continuous Fourier Integrals from which our DFTs and FFTs were derived. The integrals describe a 'function of time', i.e. something which varies with time, in terms of a set of harmonically related sine and cosine waves. These 'trig. functions' are derived from the 'exponential' term as indicated.

3.6.1. The FT & the IFT integrals

"Forward" transform
   (time to frequency)

$$S(\omega) = \int_{-\infty}^{\infty} f(t)\varepsilon^{-i\omega t}\, \partial t$$

"Inverse" transform
   (frequency to time)

$$f(t) = \int_{-\infty}^{\infty} S(\omega)\varepsilon^{i\omega t}\, \partial t$$

where:
   $S(\omega)$ is the spectrum of the time function $f(t)$
   ω is radians per unit time
   i is   (an 'imaginary' operator)
   $\varepsilon^{i\omega t}$ is   cos ωt + i sin ωt
   $\varepsilon^{in\omega t}$ is   cos $n\omega t$ + i sin $n\omega t$

   This describes how, over infinite time, an infinite number of harmonically related frquencies may be used to describe a 'function' of time.   It is perfectly feasible that we encounter 'negative frequencies' in this system.   One could argue that when you 'look back' you see the wave 'from the other side', it is 'going away' so exhibits a 'negative frequency'.   Alternately you could say that, looking back, you see the wave in a negative phase.   These two views simply describe the same phenomena.   Such arguments do not concern the mathematician - they are just mundane concepts, a negative frequency is simply, well, a negative frequency and that's an end of it.

3.6.2. The DFT & the IDFT summations

   We use a 'sum of K terms' to provide an approximation to the integral and thus can derive the underlying Discrete Transforms:

   We take 'K' readings from a 'signal' and perform the following calculations to convert to a harmonic frequency

spectrum (S).   Note that each harmonic value S(N) has
contributions from each data value T(L).

The exponential term is computed as a cosine (real) and
a sine (imaginary) component.   In the computer program, these
are either two separate arrays SR(K) and SI(K) or a 'complex'
array S(K).   For a complex array, each array element S(N) has
a real and an imaginary number part.

The discrete forward transform (DFT)

$$S(N)_{N=1}^{K/2} = \sum_{L=0}^{K-1} T(L)\, \varepsilon^{-2\pi i NL/K}$$

The discrete inverse transform (IDFT)

$$T(L)_{L=0}^{K-1} = \frac{1}{K} \sum_{N=1}^{K/2} S(N)\, \varepsilon^{2\pi i NL/K}$$

From formulae such as these, one can derive many
seemingly different computer programs (algorithms).

3.7 About FFT algorithms

To prepare the time series sample data for FFT it is
necessary to sort the values in a particular manner, selecting
groups of size K/2, K/4 and so on in a regular manner.   The
DFT can then run on subsets of the sorted data.   The
resulting spectrum is   "symmetrical"   about   either   the
fundamental   or   the   limiting   harmonic.   It   is   almost
characteristic of textbooks and FFT algorithms, that the
resulting spectral values   are not subsequently 'unshuffled'
prior to the drawing of (sometimes invalid) conclusions.
Many books present as "frequency spectra", graphs containing
what amount to negative frequencies, leaving the student in
wonder and amazement at the sheer magic of the algorithms.
Commonly, the double-sided spectra arise from analyses of the

odd numbered and even numbered data samples as virtually
independent sets. The manner of recombination of these
harmonic sets is of course, dependent upon the manner of
'shuffling' used on the source data and hence differs from one
to another algorithm.

3.7.1. Difficulties

Inadequate sampling can lead to Alias Frequencies.
Noise will usually fill a spectrum. Inadequate digitizing
can introduce additional spectral components - and the window
itself can cause havoc! Recall the pure note C sampled to
just fit a window and give a single spectral bar - the next
note, C-sharp, will sadly fill the spectrum with harmonics,
even though this note too is pure.

It was to unveil such mysteries that the many Signal
Processing courses were developed at Oxford, and it was from
them that this book was derived.

Chapter 4

The Fourier Transform Software Kit

4.1 The 'HARMANY' software kit

The computer program 'HARMANY' (HARMonic ANalYsis) was developed by the author as a teaching aid at the University of Oxford and has been in use since 1975. It was originally based on character-graphics with a VDU display to enable presentation in a Lecture Theatre. It was updated with a videographic display on the introduction of the RML 380Z Laboratory Microcomputer in 1978. Some thousands of undergraduates and postgraduate students & researchers have been introduced to FTs via Harmany. The two sets of linked software processes in Harmany correspond with the time and frequency domains of the Fourier system.

```
        DEMONSTRATION FOURIER PROCESSOR
                H A R M A N Y

        TIME SERIES            FREQ. SPECTRUM
                      SETUP
        Retrieve        R   F        set Freqs
        Opt f & noise   O   S        freq.-Series

                     FUNCTIONS
        Phase-shift     P   Q        freQ.-shift
        Digitize        D   C        Complement
        Averaging       A   B        freq. droop
        Hanning W.F.    H   E        Exp. extens
        Unit W.F.       U   U        Unit S.F.
        resKale         K   K        resKale

                     TRANSFORMS
        forward Trans.  T   I    Inverse trans.

        ?   for Instructions
        M   to define a new MACRO
        X   to eXecute current Macro
        Z.  to Zonk (terminate)        ?        ■
```

4.1   The menu of our demonstration kit

Because the software was designed strictly for teaching/learning use, the number of data points was deliberately restricted to a window of 16 samples and a spectrum of 16 harmonics. This deliberately exaggerates certain of the 'difficulties' of using the DFT technique and brings out a respect for some particular features of DFTs.

4.2 Time-series processes

4.2.1. Time domain data sources

Our software kit includes the setup of sine waves of 'any' frequency. We can call for 1/3 of a cycle in the window, for 3.14159 cycles, 1, 3, 20 or any other reasonable number of cycles in the window - with or without 'noise'.

We may retrieve stored TS data and study analytic processes.

4.2.2. Phase shift

Because the TS data is held in an array, we can arrange a circular shift (rotation) of the values to bring a particular point of the wave to the 'left' of the window. Because we use a COMPLEX DFT (sine & cosine components), the resulting transform will show the effects of phase shift.

4.2.3. Digitizing

The effects of coarse digitizing can be demonstrated.

4.2.4. Smoothing

We can use ordinary data smoothing algorithms such as the running-averages smoother or Z-derived filter algorithms on the TS data prior to using the DFT. The effect of using a TS smoother is to reduce the amplitude of the higher harmonics of analysis, resulting in a spectrum 'droop'. This can reduce noise components of the signal but can also distort the signal.

4.2.5. Window functions

The use of a window (i.e. taking N samples) can have a profound effect on the data. A repetitive wave, sampled over a precise number of 'cycles' is unaffected by windowing. When an 'unknown' wave is under analysis, it is unlikely that one 'cycle' will fit the window. Either part of the cycle is 'lost' or part of another cycle will be included. Such data is strictly non-analytic via a DFT. To circumvent the problem, we assume that the edges of the window are to blame and we use a WINDOW FUNCTION (WF).

The WF is designed to have zero amplitude at each edge (we multiply the first & last data values by zero) and to increase to unity amplitude over the main 'body' of the window. A common WF is a shifted cosine wave whose amplitude is zero for initial & final data values, and has a smooth progression to unity amplitude at the centre of the window.

The data is multiplied by the WF, producing a modified version of the data. Thus, to circumvent an ill-fit wave, we modify the wave to force a fit, but in such a manner that at least we know what is going on. Our software kit uses a form of cosine Window Function known as the Hanning Window Function. There are many other WFs out in the world of FTs.

4.2.6. Rescaling

Just that - for graphics convenience. Recall that because the FTs are linear transforms, rescaling does not affect analysis.

4.2.7. The DFT

This takes 16 data points (via the window function) and uses a conventional COMPLEX DIGITAL FOURIER TRANSFORM to produce 8 pairs of sine:cosine harmonic amplitudes.

4.3 Frequency-domain processes

4.3.1. Frequency domain data sources

We may set up harmonic series data. Odd-series (1, 3, 5...), even series (2, 4, 6...) or a full series (1..16) of harmonics. With this we use the IDFT to derive data on rectangular, triangular and other waveforms. We may type in a set of harmonic amplitudes to work with.

4.3.2. Frequency shift

The spectrum data may be shifted right by a factor of 2. This preserves waveshape whilst doubling the frequency.

4.3.3. Spectrum 'complement'

A 'curious' function. Each 'missing' harmonic is 'inserted' and each existing harmonic is decreased in proportion to its amplitude. The purpose is purely to illustrate certain 'curious' effects which can arise with harmonic series.

4.3.4. Spectrum droop

A 'Spectrum Function' which illustrates one alternative to 'smoothing'. Higher harmonic values are successively diminished.

4.3.5. Spectrum extension

An exponential function applied to the harmonic amplitudes. A multipurpose process.

4.3.6. Power spectrum

This is a composite spectrum obtained from products of the individual complex spectrum components. It can be useful in determination of the regions of a spectrum where 'wave power' is concentrated.

4.3.7. Rescaling

Just that! It permits us to adjust the harmonic

amplitudes to suit the graphical display.

4.3.8. The IDFT

This uses a normal inverse transform operating on spectrum data to produce a time series of 16 amplitude values.

4.4 Graphical displays

These include display of the time-series waveform and/or the complex (sine:cosine) or simple (power) spectral components. Positive-phase of a spectral component is indicated by a bright upright bar. Negative phase is indicated by a grey upright bar.

4.5 Software control

Interactive (command-driven) with a facility for setting a lengthy 'macro' command sequence.

4.6 Other demonstration software

This book includes graphs taken from software designed just to illustrate certain aspects of digital analysis. There are examples of windowing, sample & hold and of digitization processes.

Later in the book are examples of FFTs in action. The

supporting software uses one of the conventional FFT
algorithms with data capture and graphic display algorithms
added to permit use of a microphone, cassette tape player and
an electronic 'organ'. This enabled data capture from voice,
cathedral bells and organ, and ECG data. The basic window
frequency used is 25 Hz and 256 samples were drawn. This
choice was purely for graphical convenience, the display
having a width of 320 and the FFT algorithm being suited to a
power-of-2.

Chapter 5

On Waves and Windows

5.1 Samples of a wave

In Fig. 2.1 and Fig. 2.2 we saw examples of sine waves being sampled during a brief window in time. Neither of these waves crossed zero either at the beginning or at the end of the windowing period and neither had an integral number of cycles in the window.

This mismatch between wave and window is the most fundamental problem in work with DFTs.

Recall that a DFT assumes that the data acquired ARE of a cyclic wave, that previous and subsequent windows contain similar data. This is clearly true ONLY for a perfect fit of wave with window.

Recall also that a transform is not a FUNCTION. It does nothing TO the data, it simply provides an ALTERNATIVE DESCRIPTION OF the data.

With these ideas in mind we shall commence an investigation into the characteristics of the DFT/IDFT system.

5.2 Using pure sine waves

The first graph, Fig. 5.1, shows a sine wave of unit amplitude and unit 'frequency' (one cycle fits the window). The transform in Fig. 5.2 appears as a sine bar of 'unit' amplitude at harmonic position 1. There is no cosine component.

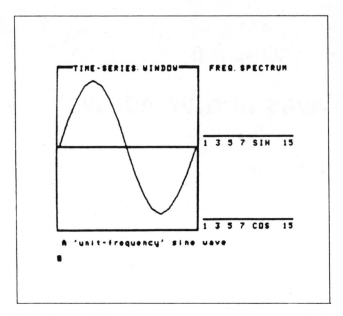

5.1   A one-cycle wave

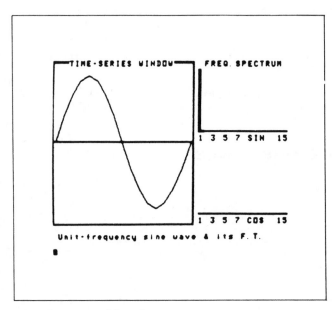

5.2   Spectrum of 1-cycle wave

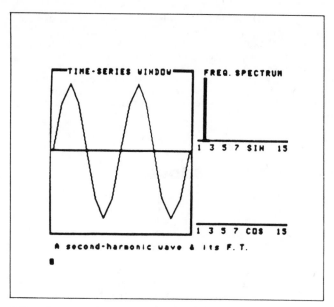

5.3   A two-cycle wave and its spectrum

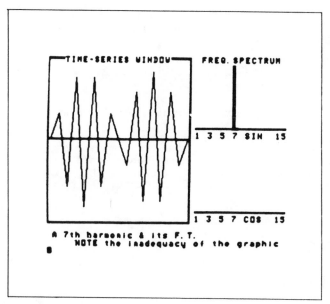

5.4   A seven-cycle wave and its spectrum

Fig. 5.3 illustrates a second harmonic sine wave and its spectrum. NOTE...The time-series graph uses straight lines to join point to point, hence the non-smooth form of the wave. We have no reliable information on how to interpolate in the general case - interpolation is possible only in rather special cases. This is in line with that other medium of data description, the polynomial. We can obtain a 'perfect fit' polynomial to a set of data (the 'colocation polynomial') But attempts to interpolate directly result in instability.

5.3 Many cycles in the window

Reminder: There are 16 samples in the window and the associated DFT can resolve only 8 complex harmonics.

Fig. 5.4 illustrates the effect of 16 samples on a sine wave of 7 cycles. The coarse line graphic indicates a curious 'wave' yet the DFT has responded correctly giving a single spectral bar at the correct position.

In Fig. 5.5 we have applied a 9-cycle wave to which the line graphic has responded by suggesting that the window contains a 7-cycle wave of negative sense. The display is an inverted form of that in Fig. 5.4. The DFT has responded to the 9-cycle wave by indicating that the harmonic number is 7. The grey bar is the device we have used to indicate a negative amplitude for this harmonic.

These examples each result in spectra having a single sine bar at the seventh harmonic position.

The first spectrum is correct - that for the nine-cycle wave is incorrect.

There are 16 samples in the window. Sampling theory shows that it is possible to analyze only half as many spectral harmonics as there are window samples. The highest harmonic of analysis in our case is number eight.

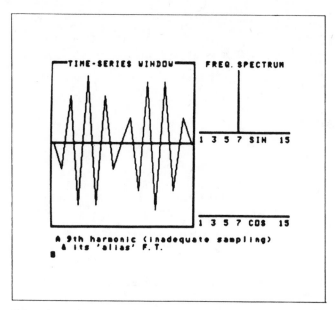

5.5   A nine-cycle wave and its spectrum

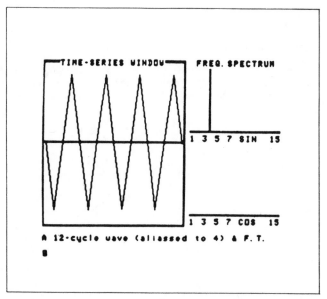

5.6   A four-cycle alias of a twelve-cycle wave

The defect in our nine-cycle wave is not in the transform - it is due to inadequate sampling.

This is so common a problem that a special name has been coined for the wrongly placed spectral lines. They are known as ALIAS FREQUENCIES - or just ALIASSES.

5.4 More aliasses

When a source wave of frequency equal to the highest harmonic of analysis is presented to the window, the result is no harmonics in the transform. This is because the samples are taken at the same points of each cycle. For an in-phase source, all the samples have zero amplitude. For an out-of-phase source, all the samples have equal amplitude, effectively a 'd.c. signal'. No variations, no frequencies.

As source wave frequency is raised further above the highest analytic harmonic, so the alias reduces until zero is reached at an input frequency of twice the analysis limit. Further increase of input frequency raises the harmonic number of the alias again.

This process iterates with the alias wandering up and down the spectrum as input frequency increases.

Again it must be emphasized that this is a defect in sampling, not a defect of transformation.

Another example of an alias is shown in Fig. 5.6 where a 12-cycle wave is subjected to a 16-sample window and then examined by the 8-harmonic analyzer. The data supplied to the DFT is that of a 4-cycle wave.

5.5 Phase-shifted waves

5.5.1. Phase-shifted sine wave

The DFT of the simple sine wave in Fig. 5.7 is itself a simple, single-component sine spectrum.

In Fig. 5.8, the wave has been phase-shifted from the left of the window. The resulting DFT has a first harmonic but with sine and cosine components. This corresponds with our findings in trigonometry. A sine wave shifted by 90 is a cosine wave. Any other phase angle may be represented as a pair of spectrum elements of amplitudes related to the phase angle.

Figs. 5.9 to 5.13 form a series of successively phase shifted waves to show the effects on the complex spectral bar.

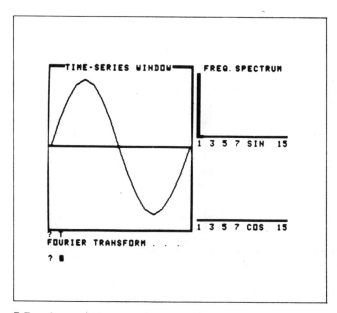

5.7   A one-cycle, zero-phase wave

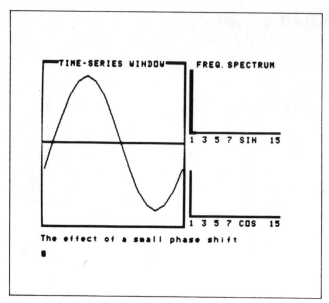

5.8 A phase-shifted version of Fig. 5.7

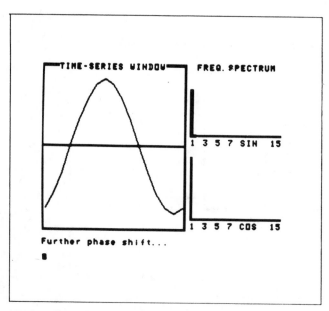

5.9 A phase-shifted version of Fig. 5.8

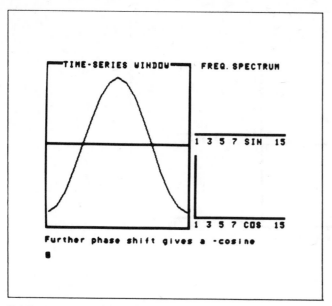

5.10 A phase-shifted version of Fig. 5.9

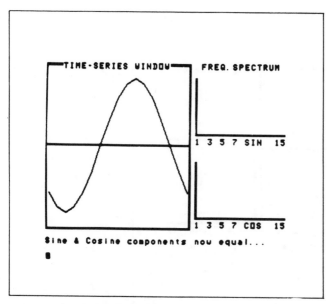

5.11 A phase-shifted version of Fig. 5.10

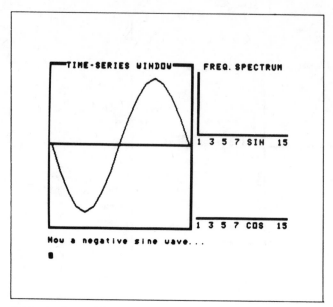

5.12 A phase-shifted version of Fig. 5.11

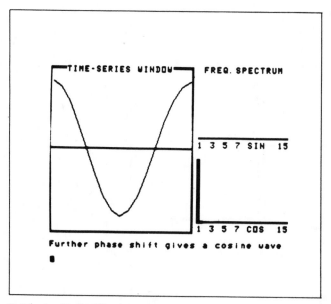

5.13 A phase-shifted version of Fig. 5.12

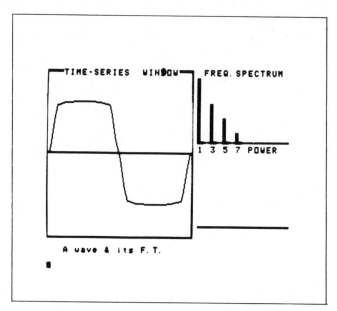

5.14 A more complicated wave

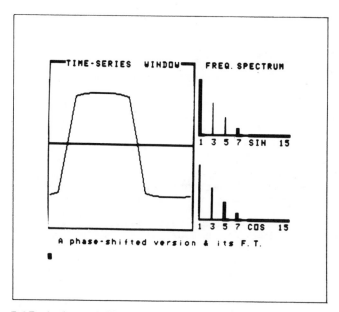

5.15 A phase-shifted version of Fig. 5.14

5.5.2. Phase-shifted rectangular wave

The next sequence is of a more-or-less rectangular wave and its DFT. The in-phase case is in Fig. 5.14 which displays a sine spectrum of four bars having diminishing amplitudes with increasing harmonic number.

In Fig. 5.15 there is a display of a phase-shifted version of the same wave. The DFT spectrum indicates the variations in spectrum bars due to the phase shift.

5.5.3. The power in a spectrum component

If we take the sum of squares of the sine and cosine components of a sine wave spectrum bar, the result is an in-phase sine component which contains all the 'power' represented in the individual sine and cosine components.

Taking this concept a stage further, we have taken the spectrum of Fig. 5.15 and formed its power spectrum. This is shown in Fig. 5.16. From this, the IDFT has given us the display of Fig. 5.17. Note however, this is a 'special case' of the successful reconstruction of a wave using a 'power' spectrum.

The 'linearity' of the DFT/IDFT transforms may be seen to extend to the 'addition' of like-numbered spectrum components and to ordinary trigonometric operations on them.

5.5.4. Phase-shifted impulse

In Fig. 5.18 we have introduced an 'impulse', a single non-zero time series value, near the start of the window. The impulse is of fundamental importance in signal processing and in control systems theory. Regular mathematical methods permit us to describe and apply impulses formally. Here we take the 'practical' case, so far as our simple DFT package is

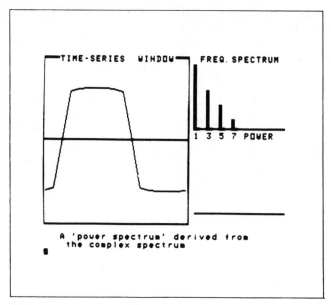

5.16 A power spectrum of Fig. 5.15

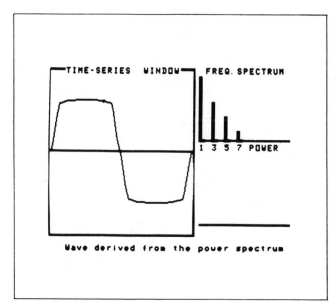

5.17 An IDFT of Fig. 5.16

concerned, and analyze the single non-zero sample which is the closest approximation we can achieve to the theoretical impulse.

Fig. 5.18 has a complex spectrum with rather interesting spectral patterns.

In successive illustrations up to Fig. 5.33, we have shifted the phase of our 'impulse' step by step and can observe the subtle variations in spectra.

Having seen the DFT in action, you may like to relate the results with the mathematical symbology of the Shifting Theorem using a standard textbook on Signal Processing.

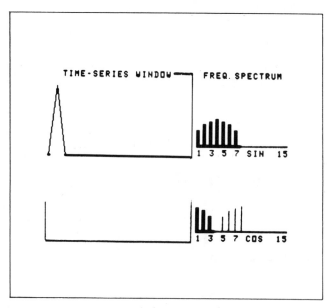

5.18 An impulse and its DFT

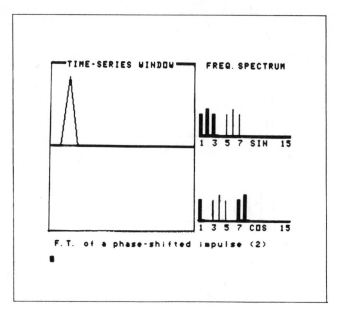

5.19 A phase-shifted version of Fig. 5.18

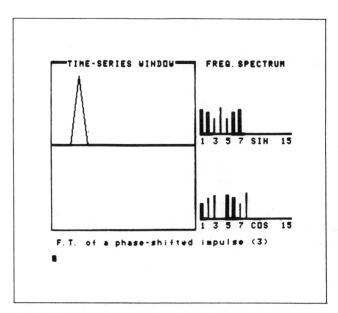

5.20 A phase-shifted version of Fig. 5.19

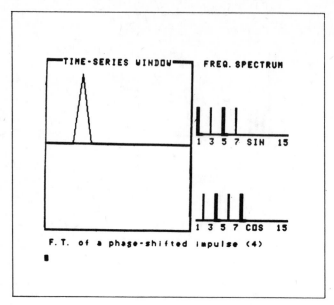

5.21 A phase-shifted version of Fig. 5.20

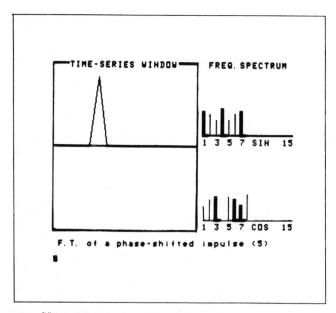

5.22 A phase-shifted version of Fig. 5.21

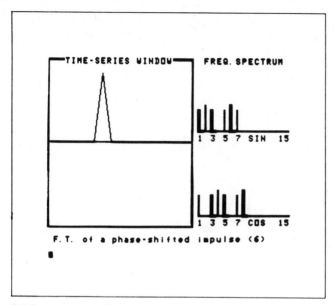

5.23   A phase-shifted version of Fig. 5.22

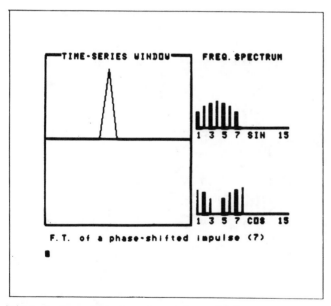

5.24   A phase-shifted version of Fig. 5.23

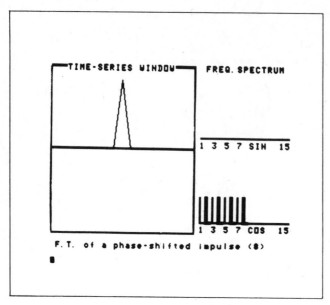

5.25 A phase-shifted version of Fig. 5.24

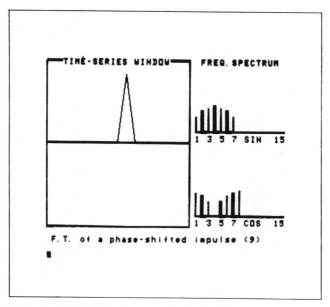

5.26 A phase-shifted version of Fig. 5.25

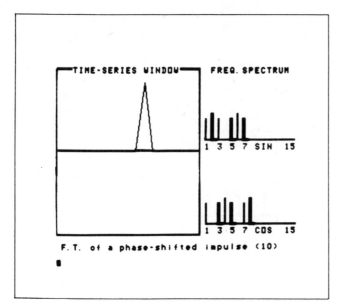

5.27 A phase-shifted version of Fig. 5.26

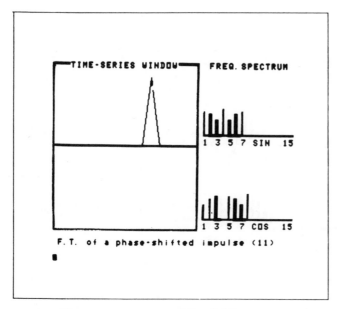

5.28 A phase-shifted version of Fig. 5.27

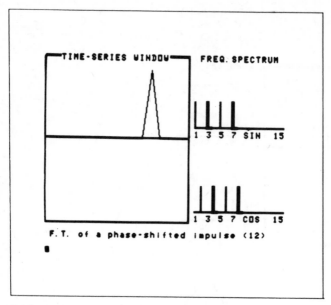

5.29 A phase-shifted version of Fig. 5.28

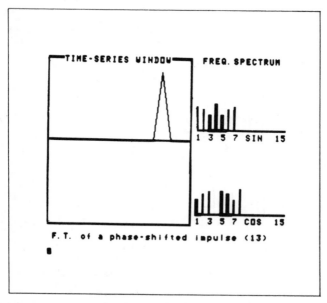

5.30 A phase-shifted version of Fig. 5.29

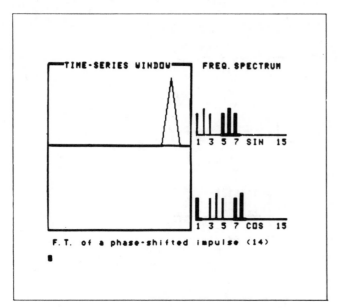

5.31 A phase-shifted version of Fig. 5.30

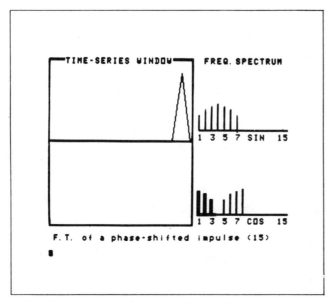

5.32 A phase-shifted version of Fig. 5.31

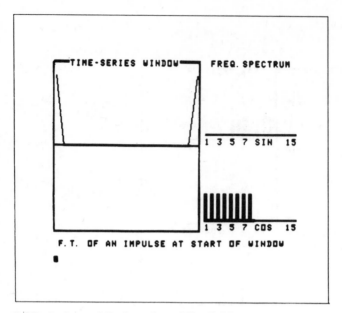

5.33 A phase-shifted version of Fig. 5.32

Chapter 6

Transforms of
Non-Integral Waves

6.1 Window misfits

In the Fig. 6.1, we have three sine cycles in the window and a simple one-bar spectrum.

The window of Fig. 6.2 contains 2.3 cycles of a sine wave. The corresponding spectrum in Fig. 6.3 displays a 'full set' of harmonics in both sine and cosine sections.

This spectrum then, does NOT contain the original FREQUENCY as a sole component. However, an inverse transform would provide a 'reconstruction' of the data taken from the window.

Again then, we see the DFT as merely an ALTERNATIVE SET OF NUMBERS representing the sampled data. The DFT has not affected this data.

However, we may not take the harmonics exhibited in the spectrum as being in any way 'real frequencies'. The original wave COULD have been composed of a single frequency or it COULD have been composed of a set of harmonically related frequencies. There is also the possibility of non-harmonically related frequencies being present in a source wave.

There could be a very large number of such sets of (analytic) frequencies. The DFT is just one such set.

The example in Fig. 6.4 used a 16-sample window containing 3.14159 cycles.

The pattern of relationship between source frequency and spectrum is beginning to emerge. There is a tendency for the spectral components to have greater amplitude at harmonic numbers close to the principal window wave frequency or frequency components.

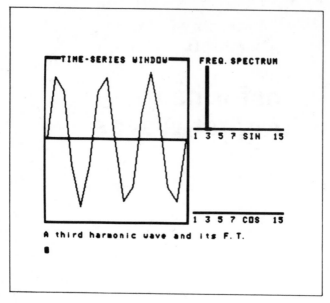

6.1 A 3-cycle wave and its DFT

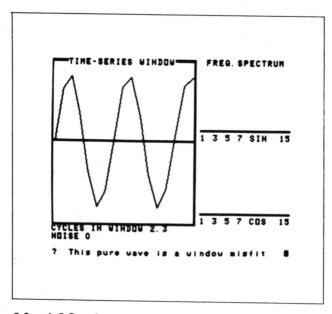

6.2 A 2.3-cycle wave

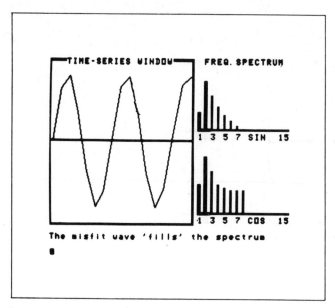

6.3 DFT of the 2.3-cycle wave

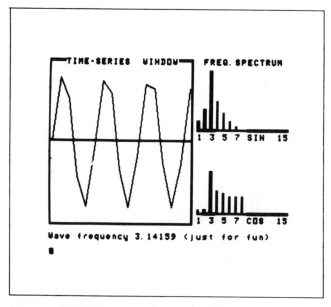

6.4 A 3.14159-cycle wave and its DFT

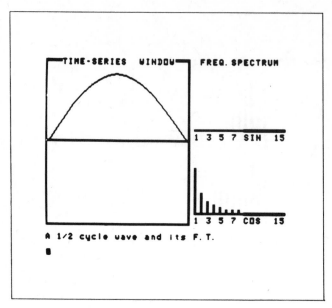

6.5   DFT of 0.5-cycle wave

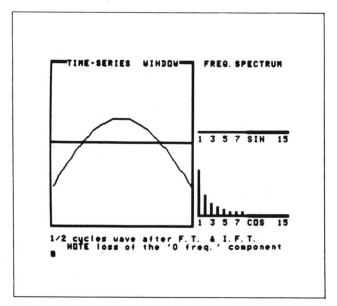

6.6   IDFT of Fig. 6.5

We see the effect of a fractional wave in Fig. 6.5. The form of symmetry of waveform suggests that the harmonic analysis would display cosine (or 'real') components. The IDFT of this wave in Fig. 6.6 again leaves the waveform unaffected save that an 'offset' or d.c. component has been removed.

We shall now examine two further examples of wave segments. Fig. 6.7 gives the spectrum resulting from a 1/3 cycle wave whilst that from a 2/3 cycle wave is illustrated in Fig. 6.8.

Clearly there is some form of aliassing going on here. The 'harmonics may be said to have 'negative frequencies' (a wierd concept). Could this be so? There are two ways to approach this. The first is to say "No, of course not, that wave could be formed from a set of (proper) harmonics - and they happen to have negative amplitudes (they are reversed in phase." The other retort is "Yes, what is wrong with a negative frequency, isn't it just an ordinary positive frequency with a reversed phase?" The graph supports either concept. A negative alias is thus quite feasible - it shows as a positive harmonic number whose amplitude is inverted.

6.2 Transform of a non-integral alias

To the DFT, the window contains the wave resulting from sampling - the wave in the sampled window, not the true source wave. Thus a source wave of 12.5 cycles is treated by our sampling system as though it were a 3.5 cycle wave, i.e. 16 - 12.5 cycles. The defect is in the sampling, not in the transformation.

6.3 Noisy waves

First we take a specimen of 'noise'. This was generated in the computer by use of a function of random

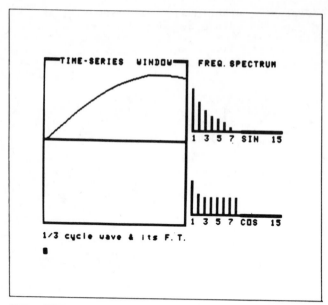

6.7 DFT of 1/3-cycle wave

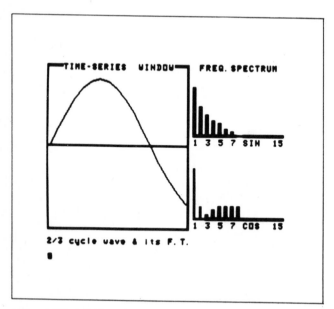

6.8 DFT of 2/3-cycle wave

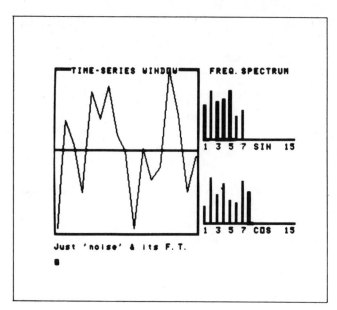

6.9 DFT of some 'noise'

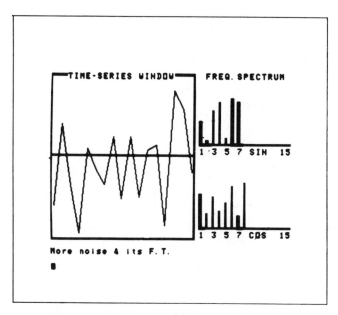

6.10 DFT of another noise specimen

numbers, so arranged as to distribute equally in positive and negative senses about a zero baseline.

The resulting DFT displays in Fig. 6.9 and Fig. 6.10 show spectra full of harmonics.

The effect of simple 'smoothing' of the time series noise data is illustrated in Fig. 6.11, where the higher harmonics of the spectrum are obviously attenuated with respect to the lower harmonics.

There is however, a clear difference between these spectra and those from the pure, non-integral wave cases.

The DFT of the pure, non-integral wave has harmonic components whose amplitudes fall at each side of the actual window-wave frequency. The noise components have seemingly random amplitudes. In practice, the noise commonly includes systematic features which can introduce considerable uncertainty into the analysis process.

The next sequence of displays uses a window having 2.3 sine cycles with noise. Fig. 6.12 shows such a wave and its DFT. In Fig. 6.13, the effect of time series data smoothing on the resulting spectrum is shown. A similar process is used on another 2.3 wave with (different) noise in Fig. 6.14 and Fig. 6.15.

6.4 Power spectrum of noise

This example is simply a spectrum of noise with no 'signal' present.

The lower display is the 'power spectrum' (product of the real & imaginary) spectral components).

A Power Spectrum of a noisy signal generally enhances the signal components in relation to the noise components.

NOTE that the taking of a Power Spectrum IS A FUNCTION which alters the 'description' afforded by either the data samples or their DFT. The inverse DFT (IDFT) of a Power

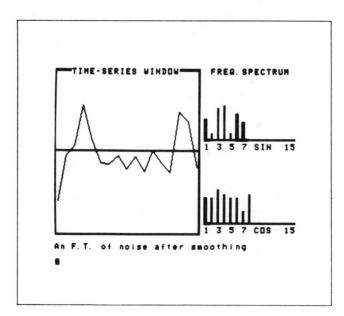

6.11 DFT of 'smoothed' noise

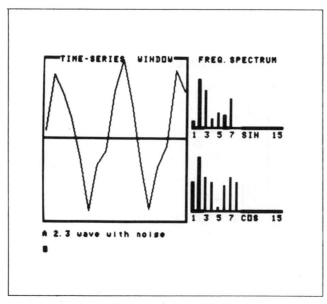

6.12 DFT of a 2.3-cycle wave with noise

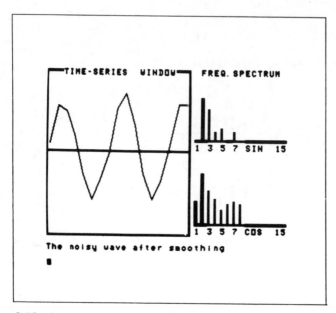

6.13   A smoothed version of Fig. 6.12

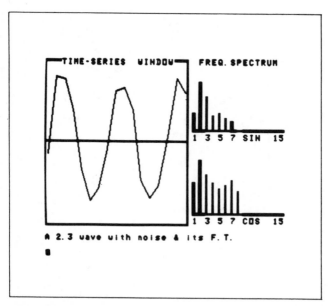

6.14   DFT of another noisy 2.3-cycle wave

Spectrum is not the same as the sampled data. Further, in practice, one finds various interpretations of the meaning of a Power Spectrum. Given that the electrical power in a circuit is proportional to the square of the voltage, one could take the square of each sample prior to the DFT. Alternatively one could take the squares of the spectral components. In the latter case, having taken the square of a Real harmonic should you simply add the square of its Imaginary counterpart - or should these component parts be left complex? A common interpretation in fact is that use of the square function enhances the difference between low-value and high-value spectral bars and thus the contribution of both Real and Imaginary parts helps identify regions of the spectrum where 'power' is concentrated. This enables design of useful filters.

Given a software kit, one should be quite clear how the designer had interpreted the meaning of 'power'.

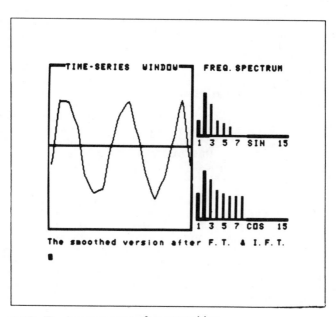

6.15 The improvement after smoothing

Chapter 7

Waveforms from Spectra

7.1 Mixed harmonic series

If we commence with a set of harmonics, we can, by use of an inverse (Digital, or Discrete Fourier Transform, obtain the corresponding waveform.

Certain "electronic organs" were once composed of sets of discs whose edges were contoured by sine waves. These contours were electrically sensed and mixed to produce the various waveforms required.

Mathematically, the exponential integral can be analyzed into an infinite number of terms which can be represented as the trigonometrical functions sine and cosine. Each angle involved contains a factor which is the particular term multiplied by the 'imaginary' value, the square root of -1. The square, or any even power of the 'imaginary operator' (root -1), is 'real' whilst any odd power of 'root -1' is 'imaginary'. In either case, the signs of the terms alternate.

In analysis of the exponential integral (Chap. 3), we can separate the even and odd terms into an even (real) series and an odd (imaginary) series. We find that the real series corresponds with a set of cosine harmonics whilst the imaginary series corresponds with a set of sine harmonics. In practice, we use a limited or 'truncated' series For efficiency, we search for series whose terms diminish rapidly. From such sets of harmonics we may produce a variety of waveforms.

It is now convenient to examine some of the characteristic features of these real (even), imaginary (odd) and complex (odd and even) series and spectra.

7.2 Odd series

Taking the odd-numbered harmonics, 1, 3, 5 etc., we can produce a variety of waveforms including a 'square' (actually rectangular) waveform.

7.3 Even series

Taking the even series 2nd, 4th etc. provides a different form of symmetry with the most 'simple' being the asymmetric triangular (sawtooth) waveform.

7.4 Full series

Taking all harmonics leads to yet another class of waveforms including the 'impulse', 'wave packet', and the 'null', 'steady-state' or zero-frequency wave.

7.5 Relative amplitudes

The 'simple cases', square, sawtooth and impulse, have simple amplitude relationships between the harmonic components as will be shown in the next few displays.

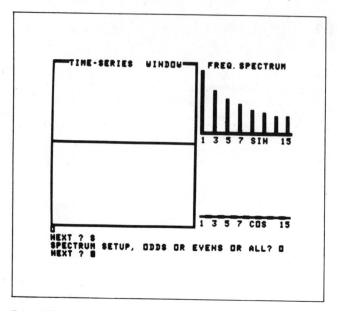

7.1 We have set up a spectrum of odd harmonics

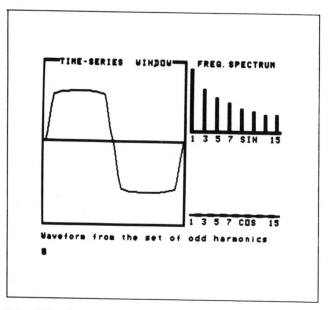

7.2 Waveform from the IDFT of Fig. 7.1

7.6 An odd series (square wave)

The graph of Fig. 7.1 shows a set of odd harmonics whose amplitudes diminish according to the harmonic number. The inverse transform in Fig. 7.2 shows an almost square wave. The coarse graphic displays accentuates the digital nature of the system. Whereas a truly square wave would rise from low to high level in zero time, our time axis has spacing whose duration is one sampling interval (one sixteenth of the window), hence the slanted rise and fall. We shall deal with the slightly rounded 'horizontals' later.

7.7 Non-uniqueness of the DFT

We commenced with a 16-component spectrum, took its IDFT and produced a time series of 16 'samples'. In the next display, Fig. 7.3, we have taken the DFT of the wave and note that there are just 8 harmonics - as we would expect from the elementary theory. The new spectrum components have different amplitude ratios from those of the original spectrum (Fig. 7.2) and yet represent the same waveform.

Thus, even among harmonic series, we see that there are alternative series from which a given waveform may be generated or by which a given waveform may be described.

The DFT then, is just one of the possible descriptors of a given wave.

A particular point to note is that a square wave has a transform which covers the spectrum.

Two further examples (Fig. 7.4 and Fig. 7.5) illustrate the effects of phase-shifted 'rectangular' waves on the DFT spectrum. The spectrum of Fig. 7.5 displays the alternations of harmonic sign mentioned under para. 7.1. However, the spectrum gives harmonic 1 as negative. This is because the spectrum was derived from an inverted waveform.

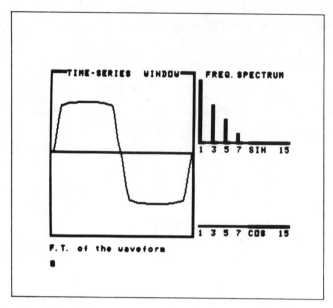

7.3   The DFT of Fig. 7.2 waveform (compare with Fig. 7.1)

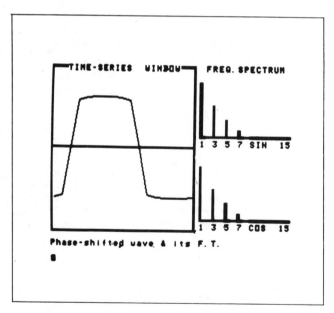

7.4   Phase-shifted version of Fig. 7.3

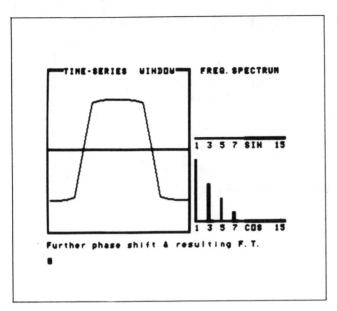

7.5   Phase-shifted version of Fig. 7.4

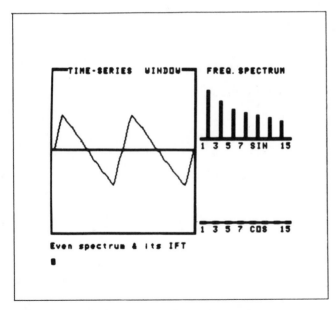

7.6   IDFT from a spectrum of even harmonics

7.8 An even series (triangular wave)

The displays of Fig. 7.6 and Fig. 7.7 show an even harmonic series, with amplitudes inversely dependent upon harmonic numbers, producing a triangular wave.

As before, we take the IDFT and its DFT to show the alternative descriptors of the same wave.

7.9 Use of an interpolating IDFT

One may be tempted to use an IDFT with less than unity steps to produce interpolates of the waveform. Sadly, this does not work in the general case. It simply introduces 'ripple' into the (now false) waveform.

The effect is similar to that when one uses a polynomial to analyze a set of equi-spaced ordinates and attempts to interpolate from the polynomial terms.

7.10 A full series

A full series of harmonics at equal amplitudes inverts to a zero wave. If phases are reversed on alternate harmonics, an impulse can result.

In Fig. 7.8 and Fig. 7.9 we see the effect of our transforms on a full set of (diminishing amplitude) harmonics. As before, we start with 16 harmonics and end with just 8 (the 8th being zero of course!).

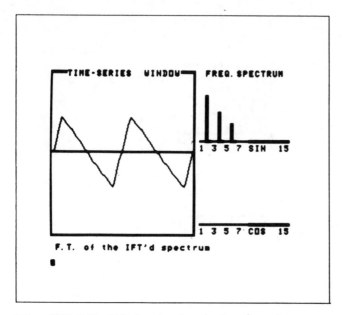

7.7   DFT of Fig. 7.6 showing foreshortened spectrum

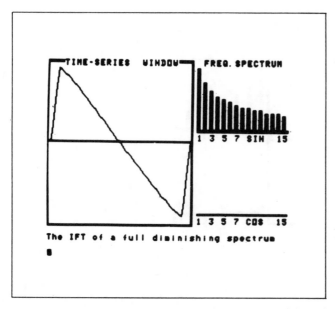

7.8   IDFT of a 'mixed' (odd, even) spectrum

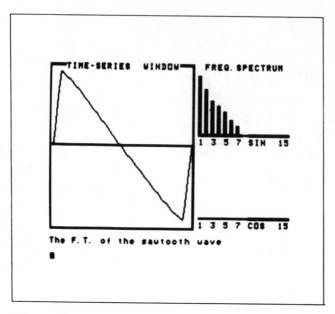

7.9   DFT of Fig. 7.8

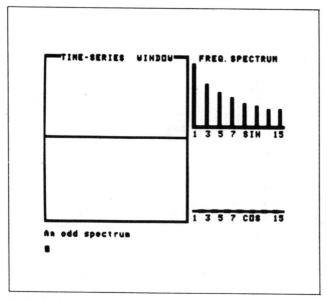

7.10 A spectrum of odd harmonics

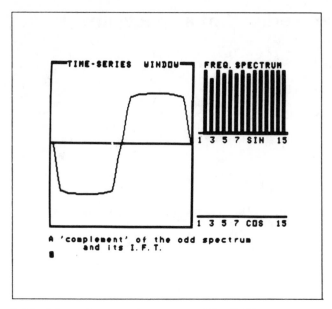

7.11 A 'complementary' spectrum of Fig. 7.10

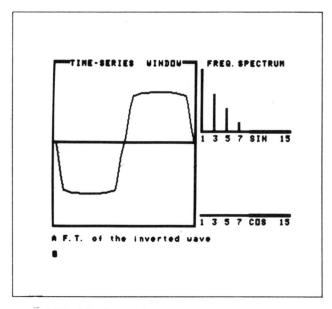

7.12 DFT of the inverted wave

7.11 The 'complement' of a spectrum

Fig. 7.10 illustrates a form of spectrum which we have met before. We now 'insert' any missing (zero-valued) harmonic, and modify any previously existing harmonic value in proportion to its amplitude. This is akin to taking the '9s' complement of a decimal number. The resulting spectrum and its inverse transform are shown in Fig. 7.11. The relationship between this 'curious' complement and the waveform is seen. The DFT of the inverted waveform is given in Fig. 7.12.

The IDFT of the 'complementary' spectrum is simply the negative (an arithmetical complement) of the waveform. The DFT of the inverted waveform returns to the normal spectrum format but with negatively signed spectrum bars.

However, it may be difficult to detect a difference between an inverted waveform and a phase-shifted version of the same waveform. Again we see that there are alternative descriptions of the effects.

Chapter 8

Window Functions

8.1 The purpose of window functions

In earlier chapters, we met numerous forms of spectrum having both sine and cosine components and a 'full' set of harmonics - 'full spectra'.

In analysis, we must recognize characteristic features associated with the spectra of 'unknown' waves.

For example, noise 'fills' a spectrum - so do square, triangular and 'misfit' waves. Harmonic analysis seeks to determine the nature of complex spectra.

The simple non-integral sine wave transforms to a 'full spectrum' - with amplitudes diminishing above and below the windowed wave 'frequency'. A wave which appears with 1.3 cycles in the window displays a spectrum which could be mistaken for that due to a triangular wave. However, as we shall see later, use of more samples and harmonics does help. With greater spectral resolution, we find that the spread of harmonics due to misfits is less than that due to waves containing steps, ramps or spikes.

Even with increased resolution, there are still very considerable difficulties in interpretation so we search for another way to improve analyses.

The pictures of Fig. 8.1 and Fig. 8.2 denote the following experiment:

Set up 1.3 cycles from a sine wave, transform, take a power spectrum and inverse.

The resulting wave is now triangular so the power spectrum was of no help in this instance.

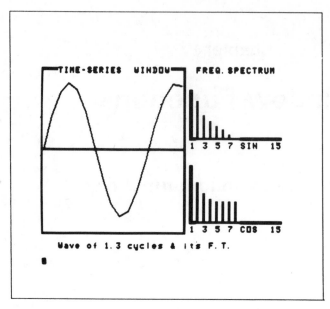

8.1 DFT of a 2.5-cycle wave

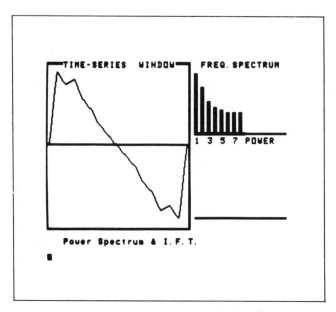

8.2 A 'power spectrum' of Fig. 8.1 and its IFT

However, the original wave was obviously 'simple' near the middle of the window but there is a 'step' (and other effects) at the edge(s) of the window. Recall that the previous and subsequent windows contain similar waveforms so that what our sampling technique 'observed' was a wave with phase transitions.

PROPOSITION...This effect is due to the 'edges' of the window. Akin to the effects of the edges of the slit in a photo-spectrometer.

If this proposition is true then

a) the window distorts the signal, effectively adding the spectrum of a transition wave, such as a step function, when it includes part of another cycle of the wave

b) we could alleviate the problem by diminishing wave amplitude towards the edges of the window. We refer to this as applying a WINDOW FUNCTION.

The Window Function (WF) will diminish the 'edge effects' of the window and reduce the tendency to introduction of spurious effects into the spectrum.

However, window functions themselves appear as spectral components having a complex but predictable nature.

NOTE...Be clear as to the difference between a WINDOW and a WINDOW FUNCTION and their individual effects. The one enables us to 'see' a segment of signal whilst the other modifies that view.

8.2 The Hanning window function

The 'smoothest' function which will present total 'insertion loss' at the edges and zero insertion loss at the centre of a window is a negative cosine wave shifted by +1 and halved in amplitude. The function is represented as a set of 'n' values, corresponding in time positions with the sample points. In use, each sample value is multiplied by its corresponding WF value prior to the transform. The Hanning

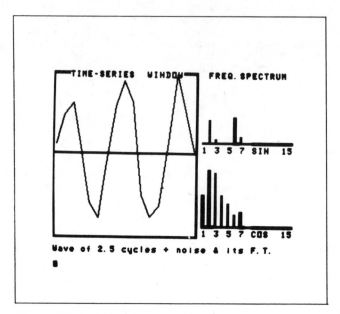

8.3　DFT of a 2.5-cycle wave with noise

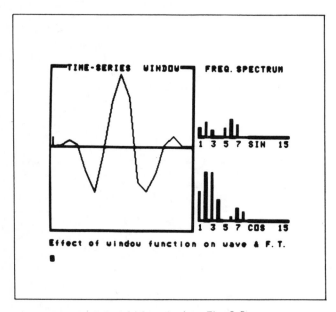

8.4　Use of a 'window function' on Fig. 8.3

Window Function is based on this concept.

The effect is illustrated in Fig. 8.3 where a misfit wave with noise is transformed and displayed.

This wave is then passed through the Window Function in Fig. 8.4 and again transformed with a clear 'improvement' in analysis.

We must now investigate what is actually happening.

8.3 Window function and a pure integral wave

The charts in Fig. 8.5 and 8.6 show the effect of the WF on a 2-cycle wave.

There are 'intermodulation products' whose existence is predicted in elementary trigonometry (the 'double-angle formulae') and which are described as 'sidebands' in AM broadcast radio. They are at harmonic positions +1 and -1 of each spectral component and their amplitude is half that of the windowed wave.

Now a final confession regarding the spectrum graphics. The amplitudes of the bars are displayed via a log. function simply to provide a compact display and not 'lose' the low valued components. This was simply for graphic convenience and does not in any way affect the course of the formal argument.

Fig. 8.7 shows the effect of our window function on a 4 cycle wave. Again there are additional spectral lines above and below the main bar.

The next displays (Fig. 8.8 and Fig. 8.9) are of a 1.5 cycle wave and the effect of the WF. These are followed by the 2.5 cycle wave via the WF in Fig. 8.10.

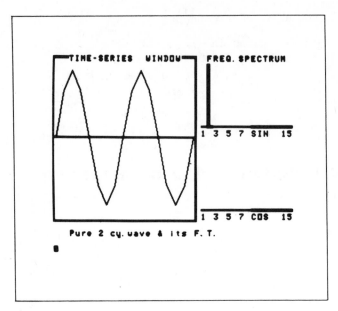

8.5 DFT of a 2-cycle wave

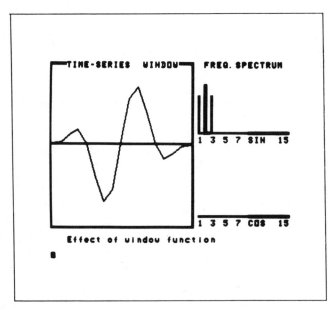

8.6 DFT of 2-cycle wave via window function

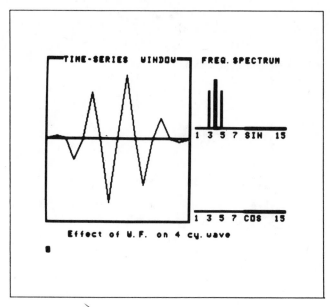

8.7   DFT of 4-cycle wave via window function

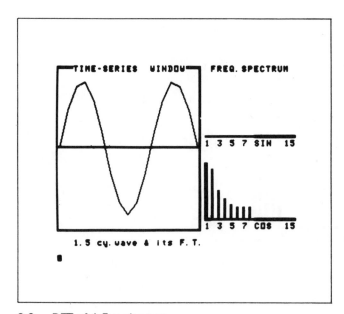

8.8   DFT of 1.5-cycle wave

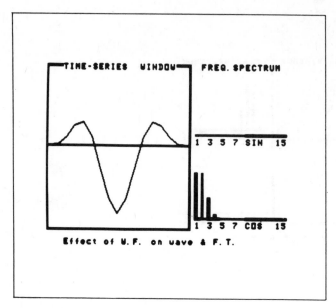

8.9   DFT of 1.5-cycle wave via window function

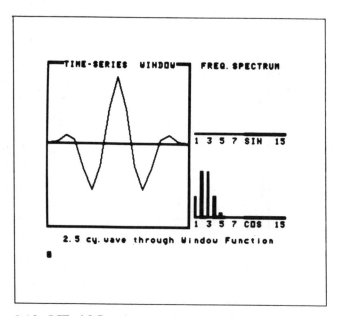

8.10 DFT of 2.5-cycle wave via window function

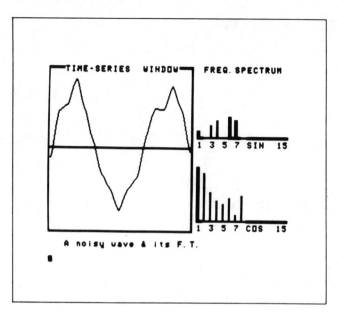

8.11 DFT of a noisy wave

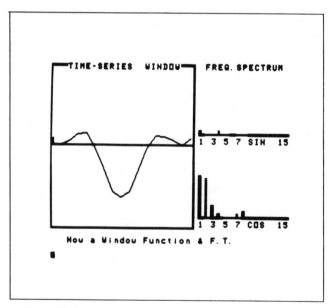

8.12 DFT of a noisy wave via window function

8.4 The effects of the wf

The WF tends to produce a peak of spectral response close to the 'true' windowed frequency whilst reducing the 'full spectrum' effects due to the window edges.

Note how the various 'sideband' harmonics have combined with 'original' harmonics to modify the resulting amplitudes.

8.5 Powers of wf spectra

The next three charts, Fig. 8.11 to Fig. 8.13 illustrate the possibility of further improvement in the analysis process when a power spectrum is taken.

The WF reduces noise to some extent, the power function gives further enhancement of 'signal' regions in relation to broadband noise.

However, it is not always possible or desirable to use power spectra. As we have seen, they can lead to extreme distortion of the analysis process.

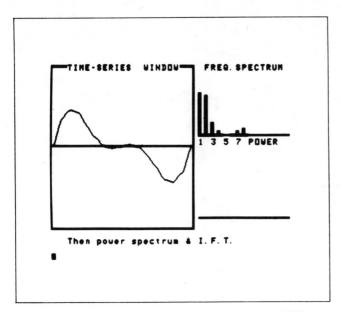

8.13 IDFT from power spectrum of noisy wave via WF

Chapter 9

The FT Kit in Practice

9.1 Interpreting spectra

In field, laboratory or clinical work, we are faced with spectra. We must learn to recognize characteristic features of the data sources from which the spectra were obtained.

We are able to 'see' these features 'through' the medium of analysis, the DFT.

Our work to date has been aimed at acquiring some understanding of the essential behaviour of the DFT in relation to component parts of signals. We now commence drawing together the threads of this work and examine the sequences of procedures commonly used in Harmonic Analysis.

9.2 Analysis of a simple but noisy non-integral wave

It may be possible to 'smooth out' the noise by using a smoothing algorithm such as

$$B(N) = (A(N-1) + A(N) + A(N+1))/3$$

or $\quad A(N) = (A(N-1) + A(N) + A(N+1))/3$

or $\quad A(N) = .25*A(N-1) + .5*A(N) + .25*A(N+1)$

etc. on the time-series data prior to the FT. Actually, although very popular, these 'smoothers' are rather poor as low pass filters and there are far better algorithms available. In passing, we note that the first of these

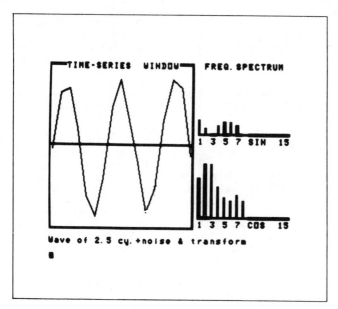

9.1   DFT of 2.5-cycle wave

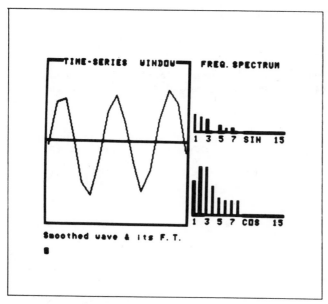

9.2   Time-series smoothed version of Fig. 9.1

algorithmic statements is 'simple' whilst the other two are recursive in that the new value of a current data item is dependent upon the previous calculation in which it was included.

Fig. 9.1 shows a noisy 2.5 cycle wave and its FT. In Fig. 9.2 we see a 'smoothed' version of the same wave and its FT. A window function has been used on the smoothed data in Fig. 9.3.

An alternative may be to perform a function on the spectrum after the FT. One possibility is to reduce the amplitudes of the higher harmonics - even to truncate those harmonics above a suitably-chosen harmonic number. This latter technique frequently results in such distortion of the transformed wave as to render analysis questionable.

The difficulty is that noise often covers the whole spectrum. One can cut harmonics which contribute nothing to the signal description but with an unknown signal, one may not know which harmonics may safely be lopped off.

Perhaps the best to hope for by noise reduction in the frequency domain is to 'roll-off' the higher harmonic amplitudes. This may well correspond with use of a suitable time-series smoother.

9.3 Frequency domain filters

A possible (and widely used) approach is to apply functions to spectral values which would correspond with the effects of regular filters such as a Butterworth or a Chebyschev filter. This can often prove to be a simple and efficient way of filtering out certain aspects of a signal - but the method is dependent for its success on detailed knowledge of the wave. One could thus class this as a 'clinical' rather than a 'laboratory' technique. It is a fine technique for routine use AFTER the wave problem has been fully analyzed.

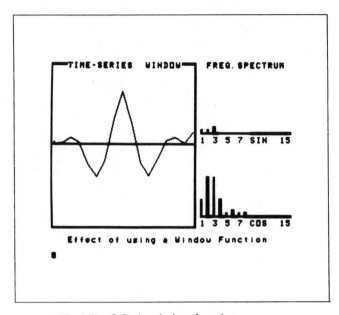

9.3 DFT of Fig. 9.2 via window function

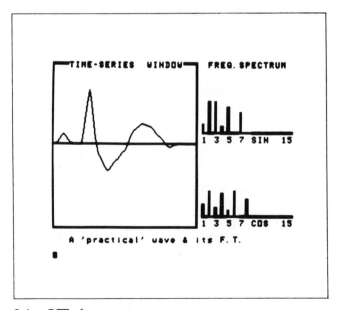

9.4 DFT of an event

9.4 Handling events

Some signals are best classified as a non-synchronous sequence of 'events' rather than as a 'continuous' signal. Each heartbeat is an event - dependent upon many factors (one of which is data on preceding heartbeats). One should not then, take a stream of ECG data and zap it through a DFT. The 'power density' of such a 'wave' is low, there is usually noise, there will be windowing effects and the constituent heartbeats are unique events. The resulting analyses may well be of little value.

If a single 'event' can be captured for analysis, its duration should be a significant fraction of the window (to ensure high power density). It should preferably be 'centred' in the window so that a window function does not cause undue distortion whilst actually reducing noise components.

The resulting harmonic numbers must then be translated back into the actual frequencies of the real world via the windowing interval. Remember that in the limit, the event is a singularity (impulse) whose DFT contains a full set of harmonics!

In Fig. 9.4 we see an 'event', spread out to fill the window and the associated DFT. A window function has been used on the same data and Fig. 9.5 shows the DFT of this. Due to loss of amplitude, we have rescaled the graph as in Fig. 9.6.

It is somewhat unclear just what conclusions could be drawn from the last three charts so, in Fig. 9.7 to Fig. 9.9 we try the effect of taking a power spectrum, noting the deleterious effect on the IDFT

So, we may conclude that the use of a short-spectrum DFT package has serious limitations. Clearly we have many tools for analysis work and they are varied in their effectivity. We must learn their individual weaknesses and strengths before we can begin to use them effectively.

A pattern is beginning to emerge, we perform some degree of signal conditioning, sample, use a window function, possibly use a time series smoother, transform and perform

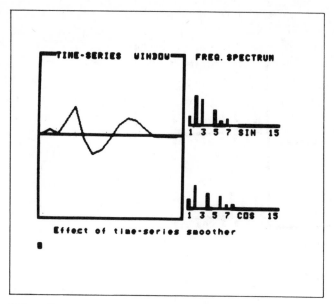

9.5　Effect of smoothing on the event

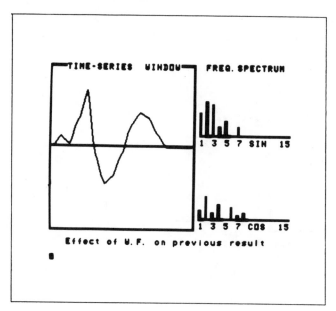

9.6　Effect of window function of Fig. 9.5

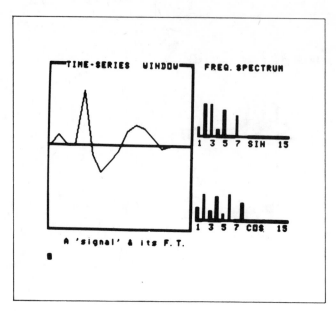

9.7   DFT of an event

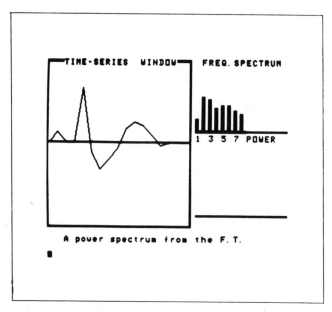

9.8   Power spectrum of Fig. 9.7

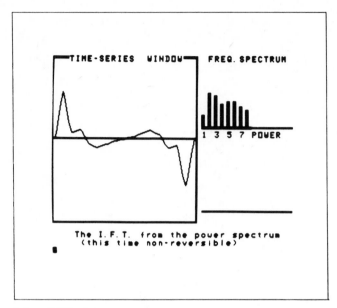

9.9   IDFT of the power spectrum

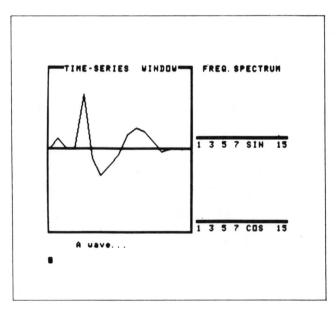

9.10   An event

carefully chosen spectrum functions. At all stages we must
visualize the effects of the chosen functions.

Events are often best handled in the Time Domain without
recourse to Harmonic Analysis at all.   Much simplification of
Time Series Analysis can occur by simply transforming ones
idea of a signal from the time domain to the domain of
wavelength   rather   than   frequency.   There is a direct
correspondence between signal wavelength and wavelength within
the window.

Again, one may consider certain types of wave to be a
sequence of unique events for which a processing algorithm is
entered in anticipation of such an event.   The adaptive
sampling algorithm devised by the Author for ECG analysis uses
just such an approach.   It provides a direct facility for
adaptation to changes in heartrate in the presence of
considerable   induced   cyclic and impulsive noise whilst
operating on ECG recordings replayed at fifty times normal
speed.   This algorithm operates on the high speed signal
directly and treats each heartbeat as an event.

To illustrate the difficulties which can arise with the
DFT technique, Fig. 9.10 and Fig. 9.11 show graphs of the
'event' used earlier but with the application of a smoother.
Even with the coarse nature of the demonstration package, it
is clear that we have lost more than noise - the signal
contour has also suffered.   This raises the question of
sampling yet again.   If we take many samples around an event,
we invoke a wide frequency bandwidth (which introduces noise)
and reduce the power density of the wave.   Clearly, we have
introduced a mismatch between our window width and the
'signal'.

9.5 Special window functions

Most books on Signal Processing list a range of WFs.
One WF rolls-off just near the edges of the window and is flat
over most of the spectrum (to minimize loss of power density
due to the simple WF).   There are many variants on the
cosine-wave and many special forms corresponding with the
analytic functions of Chebyschev, Butterworth and others.

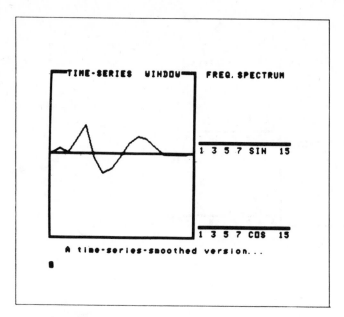

9.11 Smoothed version of Fig. 9.10

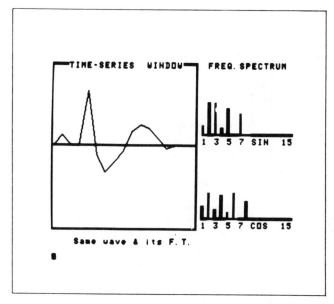

9.12 DFT of the smoothed event

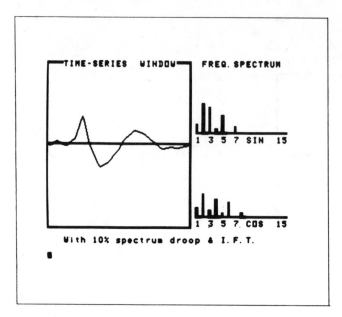

9.13 Use of 'spectrum droop'

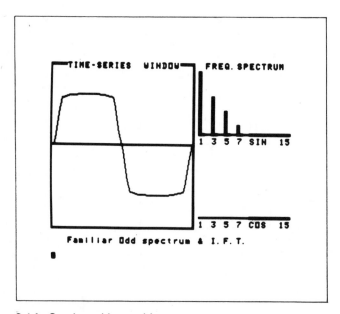

9.14 Starting with an odd spectrum

9.6 Spectrum 'extension'

NOTE... We use the terms 'harmonic number' and 'harmonic value' in the following sense: The first is the integer which 'points to' a particular harmonic, the second is the amplitude value computed (or given) to that harmonic. By use of functions on the DFT spectrum we are able to modify certain details of the (inversely transformed) source wave. One useful operation utilizes an exponential on the spectral values. An exponential constant, multiplied by an individual harmonic number is used to modify the corresponding harmonic value. This function, applied throughout the spectrum causes an exponential rise (or fall) of the harmonic values from the low to the high harmonic ends of the spectrum.

The effect is first illustrated in Fig. 9.12 and Fig 9.13 where we have taken the signal of Fig. 9.10 and used a negative spectral extension (a droop). The effect appears similar to that of the smoother used in Fig. 9.11.

Fig. 9.14 shows a waveform and its transform. Fig. 9.15 shows the result of using a positive exponent to elevate the high harmonic values, followed of course, by the IFT. Further extension can result in the 'overshoot' indicated in Fig. 9.16. This effect corresponds with excessive high-frequency gain in the time domain.

The reverse effect, exponential droop, reduces the higher harmonic values as may be seen in Fig. 9.17 and is akin to the application of a low-pass filter.

Again the process can be altered to reduce or enhance the low-numbered harmonics giving a whole range of effects.

We have put together a number of these ideas and re-examined the event discussed earlier, but this time with a different windowing so that the event occupies less of the window width. The results of a number of operations are shown in Fig. 9.18 to Fig. 9.22.

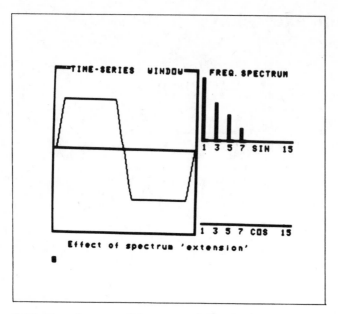

9.15 Use of exponential spectrum 'extension'

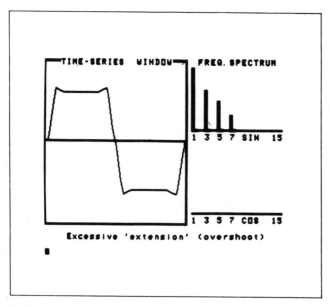

9.16 Further extension leading to 'overshoot'

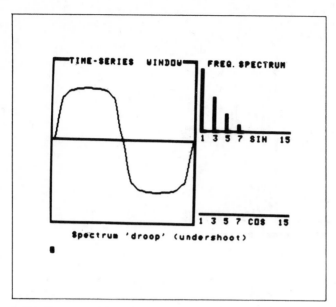

9.17 Spectrum 'droop' leads to further 'rounding'

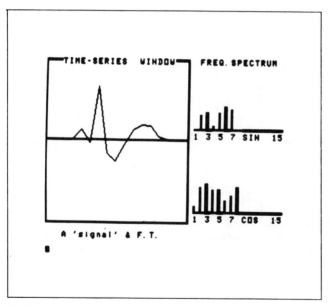

9.18 A narrower version of Fig. 9.4

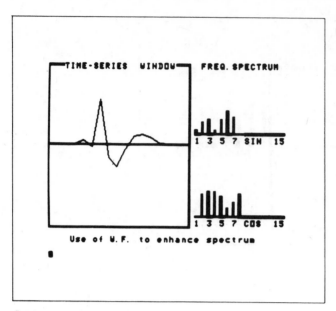

9.19 Use of a window function

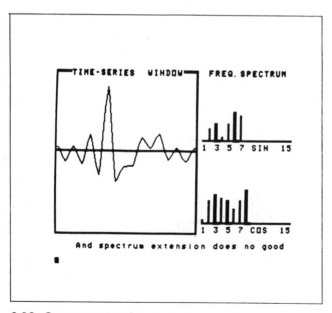

9.20 Spectrum extension spoils the waveform

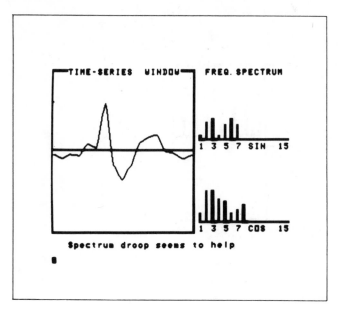

9.21 Use of spectrum droop may help

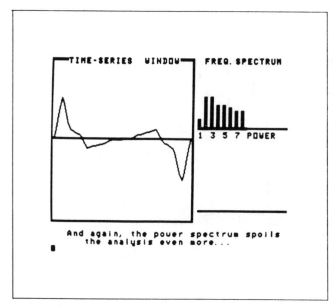

9.22 Power spectrum shows a slight 'peak'

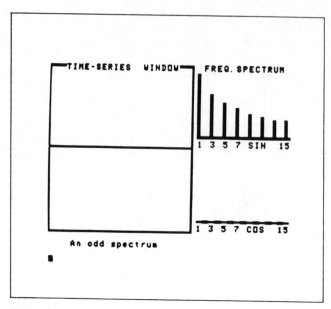

9.23 We set up that odd spectrum again

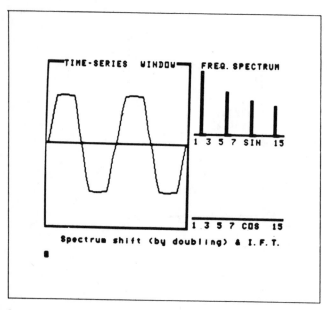

9.24 The spectral bars shifted (times 2)

9.7 Spectrum shift

A change of source frequency may be achieved by shifting the value of each harmonic from position N to position 2*N. This is illustrated in Fig. 9.23 and Fig. 9.24.

Thus, given that 'N' iterations of a wave appear in a time series window, the 'fundamental' of this will be at spectrum bar 'N'. Harmonics of the source wave will appear at positions K*N, where 'K' is the number of each source wave harmonic present.

When there is not an integral number of source wave 'cycles' in the window, each source wave harmonic will be represented as a spectrum of analytical harmonics.

Note again the distinction between the field of harmonics and 'near harmonics' in a source wave and the purely analytical harmonics, window function 'sidebands' and other effects of processing in the DFT spectrum.

Chapter 10

The Fast Fourier Transform

10.1 Why use a FFT?

The time taken to compute a DFT is considerable. The number of elementary calculations required increases enormously as the number of sample points is increased.

During the early 1970s it was not uncommon to find DFT work being undertaken on minicomputers of the day, with a single transform consuming hours of computer time.

In many applications, one had to 'tune' the transform to the signal by trying many transforms with differing values of N - many sample sizes until a clear synchronism was found. Researchers could thus work for days to obtain a single 'successful' transform.

The resolution afforded by a DFT is dependent on the number of sample points used. It is rather like an A-to-D transform where the resolution is assessed in number of bits or in the number of decimal digits. One could say that to derive 1% resolution, 100 harmonics would be required. Certainly, the difficulties highlighted by the HARMANY demonstration software used for earlier displays are alleviated (not cured) by use of many harmonics.

In practice, one needs the results in as short a time as possible - and one usually requires high spectral resolution which means many sample points.

A DFT algorithm introduced by Cooley and Tukey during the 1960s became known as the Cooley-Tukey algorithm, or more descriptively as The Fast Fourier Transform (FFT). Many versions of the algorithm have been designed and a general theory of the technique has emerged. Further, it was soon realized that the considerable similarity between the FFT and the inverse FFT could be exploited to the extent that a single algorithm can now provide both transforms.

A dual-purpose FFT algorithm usually accepts time series data to the 'real' column of its complex array and zeros to the 'imaginary' column. The FFT leaves the complex spectrum in the array. To perform the inverse transform, take the complex conjugate (reverse the sign of all 'imaginary' components) and run the algorithm again, possibly with certain minor data modifications.

10.2 The basis of the FFT

In the calulation of the sines and cosines of the many angles for the numerous harmonics, the same angular values occur repeatedly. This is to be expected as the FT relates to a harmonic series. If then, the trigonometric values for these angles were calculated once and stored, much time could be saved. Further, the quantity of the more simple arithmetic calculations could be reduced by a rearrangement of the sequence in which calculations were performed. Also, by splitting the data into smaller groups each requiring smaller (hence much faster) transforms, the computation time can be reduced still further.

In the FFT, the time-series data is rearranged systematically in a number of stages and the calculations are performed on subsets of this 'shuffled' data. This enables DFTs to be performed very rapidly on the small data groupings

The most convenient rearrangement of the data is obtained when the number of data points is a power of 2. Accordingly, the FFT is usually performed on 256, 512, 1024 etc. sample points.

This raises a sampling difficulty. Data do not usually come in conveniently sized packets. Where an experiment or a process can be sampled 'at will', it may be possible to arrange that the number of samples does suit the FFT. However it would be exceptional to find such a sampling scheme to 'tune in' to synchronism with the data.

In many cases, it is necessary to 'pad' the data with dummy values (usually zeros) to achieve the required sample size.

We must normally use a Window Function of some sort, so we pad out at the edges of the window - at the beginning and/or at the end of the available data. The WF must be designed to take account of this arrangement.

A FFT Package must include the transforms, Window Functions, means of acquiring and padding the data and of delivering the results. The transforms are usually accompanied by processor algorithms to perform such tasks as time-series data smoothing and spectral filtering. One may find 'Butterworth', 'Chebyshev' and other forms of spectrum processor as well as spectrum 'extension' and 'droop' functions. There may also be a means for 'convolution' - multiplication of the components of two spectra and other functions useful in the general application of FFTs.

10.3 Distortions in the FFT output

The FFT can impose serious distortion due to the inappropriate number of sample points, which necessitates the use of a particular type of Window Function. Clearly, the user must be cogniscent of the effects due to the chosen WF. The user must be aware of the possibility of aliasses and of the types of noise possibly present in the source data. Noise processing must be such as not to cause deterioration of the signal from the analytic point of view.

The handling of 'rogues' - data errors - is a difficult topic, both in signal processing and in statistical work. Some designers rely on use of error correction coding, but that cannot help against signal 'glitches' (static induction, lightning and other electromagnetic impulses) at the 'data source'. Popular approaches include great care to 'screen' sensors, the move to fibroptic signal guides and the use of diode signal-follower impulse 'killers'. Such circuits have been included in domestic TV sets for almost 40 years and are now conventional in radar and other communication systems.

Software models of such devices can be constructed and included as part of the 'signal conditioning' algorithms in data analyzers.

The general Fourier theory is based on time integrals from minus 'infinity' to plus 'infinity' and the spectra include 'negative frequencies'. These, according to E. Schrodinger (the founder of Wave Mechanics theory), "if they mean anything at all, probably mean the same as their positive counterparts".

Most DFT and FFT algorithms cater only for 'positive time', counting from 1..N along the time series window. Due to the initial data shuffling some algorithms deliver a curious 'folded' spectrum, symmetrical about 'zero' frequency or about the limit frequency.

Sadly, many books on Signal Processing treat this curious output as though it were a normal feature of the FT whereas it is a distortion due to the FFT algorithm. The twin spectra usually arise from the FFT treatment of odd-numbered samples and even-numbered samples as separate sets - hence two spectra. The true DFT is obtained by 'unshuffling' the spectral elements.

In the examples of FFT displays given in this chapter, 256 samples were drawn and the twin spectra are displayed. They are combined via the power spectrum algorithm.

Students should be on their guard against misinterpretation of the results. Where appropriate, the spectrum should be 'unfolded' using an algorithm based on that used for source data shuffling, prior to the drawing of conclusions. Algorithms from Mainframe Libraries should be checked to ensure that the user does understand precisely how the input data must be assembled and how the output data are to be handled. When using FORTRAN libraries, the transfer of data between a calling program and a subroutine must be organized with great care. The two principal techniques are transfer by parameter list and transfer via COMMON storage. Most library routines must use parameter list data coupling to achieve variable-sized array transfer. The most reliable form of parameter list transfer occurs when the array names occupy the same positions in the data specifications as in the parameter lists. In this connection, use of REAL declaration is preferable to use of DIMENSION. Use of COMPLEX is also quite 'safe'. Experience has shown that some compilers are liable to do unexpected things when using DIMENSION and when the sequence of declarations is different from the sequence of parameter list entries.

In taking FFT algorithms from books, it is best not to

use a version for which there is not a detailed description of the nature of the resulting output data.

The sequence of Fig. 10.1 to Fig. 10.12 shows FFTs of: voice (vowels A, E, I, O and U), cathedral bells and organ, electronic violin, electronic flute and three examples of ECG. The time-series graphs have the real spectrum above and the imaginary below with a power spectrum at the bottom. The 'raw' FFT spectra were displayed to show the 'normal' output from a typical FFT algorithm. They include no signal conditioning, no smoothing, no window function and no spectrum functions.

These displays indicate a considerable improvement over the short spectrum DFTs but, with the various aspects of the system in mind, we are better able to visualize the various effects. In particular, we note the clear spectral line of the ECG timing marker in Fig. 10.11.

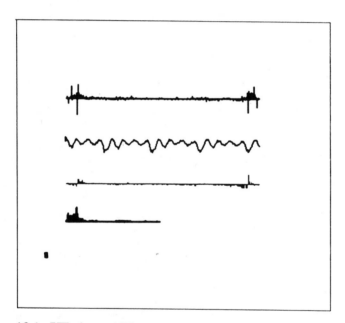

10.1   FFT of vowel 'A'

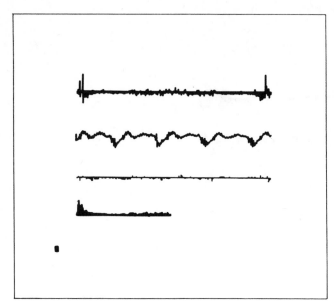

10.2   FFT of vowel 'E'

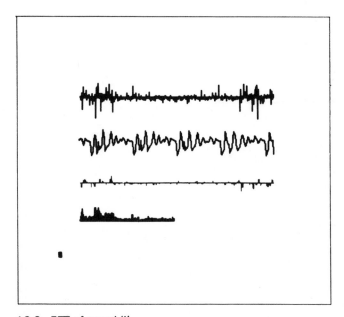

10.3   FFT of vowel 'I'

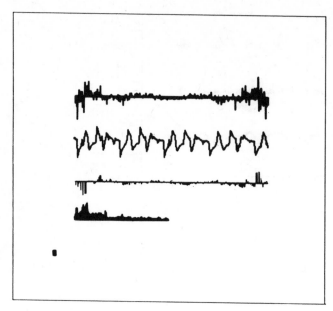

10.4   FFT of vowel 'O'

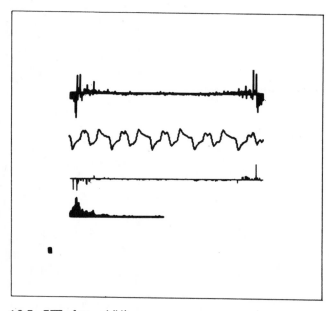

10.5   FFT of vowel 'U'

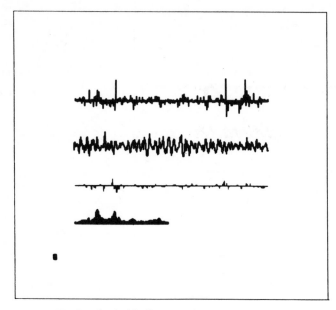

10.6 FFT of cathedral bells

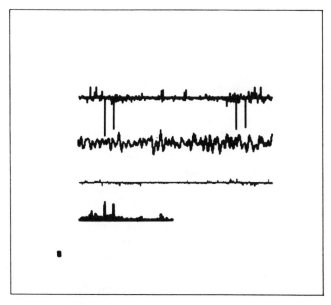

10.7 FFT of cathedral organ

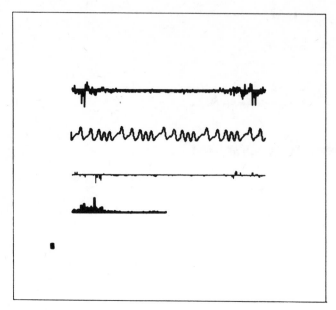

10.8 FFT of 'electronic violin'

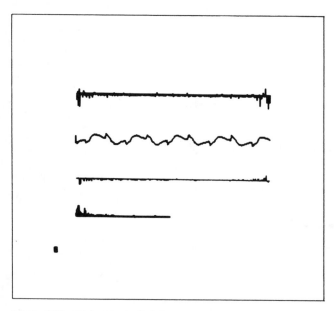

10.9 FFT of 'electronic flute'

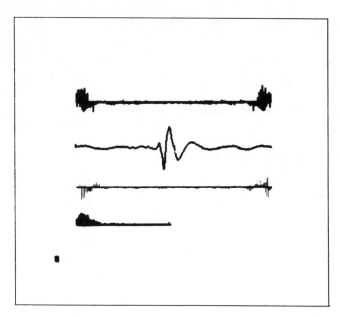

10.10 FFT of (high-speed) heartbeat

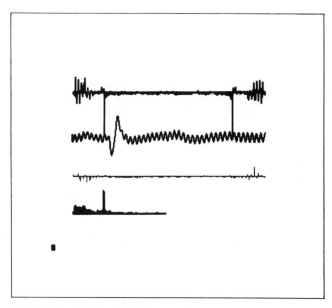

10.11 FFT of ECG with timing marker

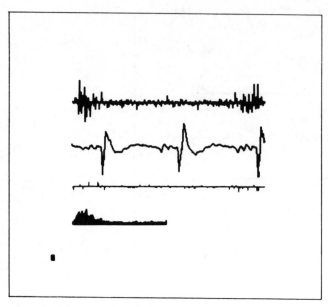

10.12 FFT of ECG with higher heartrate

10.4 On the nature of the DFT

The DFT strictly applies to a window on a signal. The theory assumes that every time window contains the same data - that the signal is STATIONARY.

The DFT is not then, suited to the analysis of ECG data because, although adjacent heartbeats are similar, each is an individual EVENT initiated by CURRENT neurophysiological conditions.

When an 'event' occupies a small proportion of the window, the Power Density of the signal is low and results may well be swamped by transform 'errors' and noise. Event analysis is not the forte of the FFT.

10.4.1. Analyzer or processor?

As a laboratory instrument, the FFT package can perform a variety of useful analytic functions - but the interpretation of spectra is fraught with difficulties of mis-interpretation. As a data processor (such as a clinical instrument), the distortions due to the FFT process can often far outweigh any presumed advantage because of the strict mathematical basis of the technique. The designer of a data processing package must have studied all facets of the task from the data acquisition process to the required display - and ensured that at every stage, the operations performed are valid in every sense.

Chapter 11

Applications for FFTs

11.1 FFTs in action

Modern FFT packages operate with comforting speed, often a thousand times faster than their more simple DFT counterparts.

It thus becomes reasonable to search for other applications of this highly successful algorithm.

11.1.1. Signal processing

It is tempting to provide a processing suite which will accept TS data, transform, filter and revert to TS for onward transmission.

There are (limited) application areas for such activities - but the distortions within the FFT algorithms (Windowing etc.) can result in a complex and sometimes ineffective system.

Noting that the DFT occupies a position in the class of 'Z-transforms', it may frequently be preferable to use some other 'Z' rather than a Fourier transform so as to design within a far more simple and efficient framework. The mathematics of the Z-transform may appear less coordinated than does the application of the FFT package, but modern Z-transform design packages such as the INTEL Signal Processing Applications Compiler (SPAC) package alleviate this problem. The conventional Z-transform approach takes a 'jump' from the (rectangular) S-plane to the (circular) Z-plane. In fact the transition may be gradual, i.e. something between the vertical S-axis and the 'unit circle' of the Z. It is then possible to design with this (relatively new) finding in mind.

11.1.2. Voiceprints

One of the attractions is that a FFT can provide a good spectrum rapidly. However, for identification, there must be a correlation between the new and the stored voiceprints.

One disadvantage of taking DFT voice spectra is that they result in large sets of data for subsequent correlation. However, this can be offset to some extent by combining the use of the FFT processor for both transformation and correlation. Another disadvantage is that, being based on frequencies only, they lose temporal information. A reversal of sounds ('IT' or 'TI' for example) would be difficult to detect in a spectrum.

The use of a pair of analog filters to separate high and low frequency bands followed by zero crossing 'counters' can provide a two dimensional (non-harmonic) spectrum very rapidly. This too, loses temporal data and must be followed by a process of correlation. It is questionable whether the DFT offers any real advantage unless embedded in high-speed logic hardware.

On an associated theme, the FFT is widely used in the analysis of the vocal tract. Using a FFT kit, it has become possible to trace physiological defects by comparing the effects of the voice 'formant' cavities via their FFTs with sets of stored records.

11.1.3. Digital filter banks

The FFT can be adapted for use as a filter bank by simply taking a set of regions from a spectrum. Simple truncation will usually result in unacceptable distortion of the results, so a carefully-chosen filter profile (Chebyshev etc.) must be used. It is usually preferable to use either a Z-transform derivation or a direct TS device such as a Resonator Bank (which operates in the Wavelength Domain rather than in the Frequency Domain).

11.1.4. Correlation

The process of data correlation, auto-correlation (comparing a data stream with shifted versions of itself) and cross-correlation (comparing two data streams) is a time-hungry process involving a search for periodic features of the data.

The correlation algorithms involve vast quantities of arithmetic operations as the data streams are cycled through the basic 'length' of the correlation.

The features of correlation involve a fundamental 'length' (like a DFT 'window') and the search for 'frequencies' (as in the use of DFTs).

In practice, one may use the FFT package for correlation in the following manner. First produce FFTs of each data stream, next undertake a (complex) product of the two spectra (harmonic by harmonic) then IFFT the product spectrum. The result can provide a correllogram.

11.1.5. The FFT in geology

As a single example, consider a gravimetric survey from the air. A traverse grid is 'laid down' for the aircraft to fly. Gravitometer output is recorded (often as an analog signal stream). Taking a single track of the air traverse, one may obtain (by sampling), a stream of numbers representing gravity measurements at a given height.

Mass concentrations below ground will affect the stream of aerial readings so as to indicate a curve centred over the 'mascon'. Take the DFT and use an exponential spectrum extension function (increasing the higher harmonics). The effect on the IDFT is that the 'hump' due to the mascon has a smaller radius - as though it were taken at a lower altitude.

By 'scaling' the spectrum extension to match the 'pitch' of the samples (horizontal distance), one can produce a series of extended spectra which would correspond with gravimetric sweeps at successively lower (including 'negative') altitudes. When the 'altitude' corresponds with the centre of mass of the underground 'object', the IDFT chart will exhibit instability - an 'oscillation'. The degree of spectral extension needed to induce this effect is a measure of the depth of the 'mascon'.

11.1.6. Surface transforms

The FFT kit makes possible the extraction of analytic spectra relating to 2-dimensional data. It is a simple matter to perform FFTs on sets of data and to treat these as orthogonal transforms - you do the groundwork, the computer does the number crunching.

The FFT has transformed signal processing by making possible, analyses which were previously just pipe dreams. The FFT deserves to be used well and wisely, but like any other tool, its limitations must be respected.

One of the alternative forms of DFT operates using Integers, and thus it offers some speed advantage dependent on the type of computer available.

Great care is needed in the interpretation of effects. There are descriptions in current applied optics literature of the 'Fourier Transform Properties of a Lens'. The effects so described are certainly space-frequency transforms, but they are continuous, non-harmonic effects and occur where the light strikes the 'edges' of components, in the 'object' e.g. 'marks' on a slide. The lens merely focusses the effect in a convenient manner. A Fourier Transform may be used to describe the effect, a FFT may be used to compute the effect. The Fourier Transform however, is quite independent of the events in the optical path, and is certainly not a 'function' of the lens.

11.2 Programming languages for FFTs

Whilst it is possible to prepare DFT packages in many

'languages', the most common choice has been FORTRAN. So many of the features of that system match the requirements of signal processing. There are high-speed COMPLEX number functions and fast array access between program segments. For 'image processing', FORTRAN permits the respecification of array space 'on the fly' in that it is often convenient to capture the (rectangular) data in a vector, to access the data as a single line segment in another vector, e.g. for a FFT, and to re-access the data as a rectangular array for graphics processing. With such techniques, one gains the high speed of FORTRAN whilst having the convenience of 2-D or 3-D array working where desired. Such forms of re-access via COMMONing are effectively zero-time data transfers.

Clearly, use of 'assembler code' offers the greatest degree of optimization but there is a gulf between the achievement of an algorithm and of a complete software package.

BASIC has been used and algorithms have been published but its interpreters take 'for ever' to perform transformations. Some variants of PASCAL are attractive, but it (like BASIC) lacks the high-speed COMPLEX functions of FORTRAN, and it to can be disappointingly slow in execution speed.

ALGOL-68 is attractive for this work but few of us have access to its compilers and libraries.

Published algorithms vary greatly between their claims and their achievements. When using such algorithms, take great care to ensure that the variant of the 'language' is compatible with 'your' compiler. A common source of difficulty is that the lower bound of arrays in some languages and compilers can be zero and in others it must be 1. Translation of algorithms from one form to the other can be a rather painstaking affair.

The data shuffling operations at input and output of the FFT can diminish its speed advantages, the turnover point between DFT and FFT is often around size 64 to 128 points. For smaller window sizes, the ability to select any window size rather than just a power of 2 may favour the DFT for speed and simplicity.

The scientific software libraries (such as NAG) provide FFT facilities. Each must be examined carefully before use - to ensure that your mode of use and interpretation of results is correct.

Finally, we should enquire whether the computer and its various 'languages' do provide the best approach to such work. Multi-parallelism can assist in certain cases, especially where task-specific hardware can be constructed. But is our mathematics yet suited to the analysis of complex signals embedded in noise? You may like to ponder on the feasibility of producing synthetic neuro-like networks operating in the probability domain with a whole-image sensor coupled to a whole-image processor where even correlation operates 'in parallel' between vast numbers of stored images (Reference 5).

Appendix

1. A reminder about complex numbers

Each complex number has two parts, a 'Real' part and an 'Imaginary' part which may not be added etc. arithmetically. The complex number is usually given by:

```
        Engineers    Mathematicians    FORTRAN
as
                                       REAL A, B
                                       COMPLEX Z
        z = a + jb   z = a + ib        Z = COMPLX(A, B)
```

where: the a's and b's are ordinary (mixed) numbers
       A and B are FORTRAN REAL numbers
       a and A are the Real parts
       b and B are the Imaginary parts
       z and Z are complex numbers
       i and j are root -1 (the 'imaginary' operator)

A COMPLEX value could look like:

z = 1 + i2   or in FORTRAN:

Z = (1.0, 2.0)

The variable 'a' (or A) is thought of as being on the X-axis whilst 'b' (or B) is on the Y-axis. Thus the 'length' (usually called the radius) of Z is:

R = SQRT(A**2 + B**2)

Note... the use of B*B or B**2 is more accurate than the real power, B**2.0 due to the method of evaluation in a computer.

The corresponding FORTRAN function is:

R = CABS(A, B)

The direction (angle) of the vector Z is:
```
        THETA = ATAN(B/A)
or      THETA = ATAN2(B,A)
        A = R*COS(THETA)
        B = R*SIN(THETA)
```

This is where the angles and angular functions 'suddenly' appear from, it is in the expression of the exponentials in the formulae via the Argand Diagram (the X,Y coordinates scheme).

2. Some notes on FORTRAN-80

If you have a FORTRAN-80

or another compiler/library which has no type COMPLEX, you can still undertake complex number work as follows:

```
C   Use separate arrays for Real and Imaginary parts
        REAL X(256), Y(256)
C   and don't forget to clear, read, write each of them
Careful: Don't use 'AREAL' & 'AIMAG' as names, FORTRAN will
Create quite a fuss about them.
C     FORTRAN will also fuss over BREAL etc.
C     but it won't mind BIMAG (it has a function AIMAG)
C
C   Use pairs of REALs such as:
        REAL AR,AI, BR,BI, CR,CI
C
Complex addition:
        AR = BR + CR
        AI = BI + CI
C
Complex multiplication:
        AR = BR*CR - BI*CI
        AI = BR*CI + BI*CR
C
Conjugate of (AR, AI) is (AR, -AI)
C And when you sort remember to exchange both Parts
```

Be careful also with 'F-80' to ensure no 'mixed arithmetic' such as A*4 (which should be presented as A*4.0). Similarly, A*N should be A*FLOAT(N).

When reading from an ADC (analog to digital converter) port, use type INTEGER (not INTEGER*1 or type BYTE) and note that the number which arrives in memory is signed (whether the ADC is set to bipolar or unipolar inputs). Further, if your ADC is 10 or 12 bits and you need to read the two separate bytes, each byte will arrive as a sign-extended integer. If the value of the (unsigned) byte is greater than 127 the resulting integer will be 'negative'. I.E. the lower byte may have a true value of 223 but, because the leftmost bit is 1, FORTRAN 'extends' this 'sign bit' through the entire upper byte of the integer. The pattern of bits then 'looks like' the value -32. By adding 1 to the upper byte (add 256 to the INTEGER), the upper byte is cleared, and the integer sign is now positive leaving the proper value for the whole integer. Note that if a BYTE is used to receive the ADC byte, any arithmetic will be undertaken in INTEGER and the sign extension will be applied as before. This is one of those 'sad' cases where we must force a sign change by causing an overflow. The 'normal' approach would be to use the ABS function - but that would result in a wrong value.

Some may say that this is 'bad' computing practice. However, 'correct' programming would cause the compiled program to give the wrong answer. If you care (or dare) to alter the compiler to treat a port as a bit pattern rather than as a signed number, then you may use 'correct' programming. Alternatively, write an assembler coded port-handler routine. The trap which awaits you then is that many F-80 (plus computer plus I/O) systems will not permit direct (PEEK) access to memory locations above 32K. You may find it necessary to transfer from port to a low location and to PEEK the variable from there. To do this, 'save' that location in a variable), move the data from the ADC port via an assembly routine to the (saved) location, then PEEK it into the required variable (usually in an array element) and finally restore the original value of the ill-used transfer location. As you see, this is one of those 'grey areas' of computing which we would have hoped to be free from in modern software tools, but most languages are supported by software which was not designed to serve every possible usage. In these cases, it is sometimes difficult to 'see' what a program is actually doing and to understand why certain techniques are adopted. It is a case of knowing what one wishes to do and of finding a way around a seemingly intractable difficulty.

References

THEORY AND APPLICATION OF DIGITAL SIGNAL PROCESSING

    L. R. Rabiner & B. Gold, Prentice-Hall, 1975

SIGNALS, SYSTEMS AND THE COMPUTER

    Paul M. Chirlian, IEP, 1973

INTRODUCTION TO DIGITAL FILTERS

    T.J. Terrell, Macmillan Press, 1980

PROGRAMS FOR DIGITAL SIGNAL PROCESSING

    IEEE Acoustics, Speech & Signal Processing Committee, IEEE Press, 1979

POST-DIGITAL ELECTRONICS

    F.R. Pettit, Horwood, 1982

CB Student Text Series

Cobol for Mainframe and Micro, Watson D, 178pp, 1985, ISBN 0-86238-082-0

Comparative Languages, Malone Dr J R, 129pp, 1984, ISBN 0-86238-067-7

Compiler Physiology, Farmer M, 205pp, 1985, ISBN 0-86238-064-2

Computer Systems: Where Hardware Meets Software, Machin Dr C H C, 1986, ISBN 0-86238-075-8

Databases and Database Systems, Oxborrow Dr E A, 1986, ISBN 0-86238-091-X

File Structure and Design, Cunningham M, 212pp, 1985, ISBN 0-86238-065-0

Fortran 77 for non-scientists, Adman P, 109pp, 1984, ISBN 0-86238-074-X

Fortran 77 Solutions to non-scientific problems, Adman P, 155pp, 1985, ISBN 0-86238-087-1

Fourier Transforms in Action, Pettit F, 136pp, 1985, ISBN 0-86238-088-X

Generalised Coordinates, Chambers Dr L G, 104pp, 1985, ISBN 0-86238-079-0

The Intensive Pascal Course, Farmer M, 111pp, 1984, ISBN 0-86238-063-4

An Introductory Course in Computer Graphics, Kingslake Dr R, 144pp, 1984, ISBN 0-86238-073-1

Linear Programming, Lau Dr K K, 115pp, 1984, ISBN 0-86238-071-5

Operating Systems through Unix, Emery Prof. G, 100pp, 1985, ISBN 0-86238-086-3

Programming Language Semantics: An Introduction, Rattray C, 1986 ISBN 0-86238-066-9

Computer Science (general)

Communication Network Protocols, 2nd Ed. Marsden Dr B W, 352pp, 1986, ISBN 0-86238-072-3

Computer Networks: Fundamentals and Practice, Bacon, Stokes, Bacon, 109pp, 1984, ISBN 0-86238-029-4

Computers from First Principles, Brown M, 128pp, 1982, ISBN 0-86238-027-8

Database Analysis and Design: An Undergraduate Text, Robinson H, 375pp, 1981, ISBN 0-86238-018-9

Data Bases and Data Models, Sundgren B, 134pp, 1985, ISBN 0-86238-031-6

Discrete-Events Simulations Models in Pascal/MT+ on a Microcomputer, Jennergren L P, 135pp, 1984, ISBN 0-86238-053-7

The Essentials of Numerical Computation, Bartholomew-Biggs M, 241pp, 1982, ISBN 0-86238-029-4

Fundamentals of Microprocessor Systems, Witting P A, 525pp, 1984, ISBN 0-86238-030-8

Information Ergonomics, Ivegard T, 228pp, 1982, ISBN 0-86238-032-4

Information Modelling, Bubenko J A, 687pp, 1983, ISBN 0-86238-006-5

Information Technology and a New International Order, Becker J, 120pp, 1984, ISBN 0-86238-043-X

Programmable Control Systems: An Introduction, Johannesson G, 136pp, 1985, ISBN 0-86238-046-4

JSP:A Practical Method of Program Design, Ingevaldsson L, 194pp, 1979, ISBN 0-86238-011-1

Proceedings of the Third Scandinavian Conference on Image Analysis, IAPR, 426pp 1983, ISBN 0-86238-039-1

Simula Begin, Birtwhistle, Dahl, Myhrhaug, Nygaard, 391pp, 1979, ISBN 0-86238-009-X

Systems Programming with JSP, Sanden Dr B, 188pp, 1985, ISBN 0-86238-054-5

Teaching of Statistics in the Computer Age, Rade L, 248pp, H/b, 1985, ISBN 0-86238-090-1

Techniques of Interactive Computer Graphics, Boyd A, 240pp, 1985, ISBN 0-86238-024-3

Transnational Data Flows in the Information Age, Hamelink C J, 115pp, 1984, ISBN 0-86238-042-1